Practical Manual of Quality Function Deployment

Davide Maritan

Practical Manual of Quality Function Deployment

Springer

Davide Maritan
Soave
Verona
Italy

ISBN 978-3-319-08520-3 ISBN 978-3-319-08521-0 (eBook)
DOI 10.1007/978-3-319-08521-0

Library of Congress Control Number: 2014948074

Springer Cham Heidelberg New York Dordrecht London

Printed on acid-free paper

Springer is part of Springer Science+Business Media (www.springer.com)

To Monica,
Sebastiano and Anna,
my life

Foreword

The importance of new product development (NPD) for a company's growth and prosperity is widely recognized. Faced with an increasingly volatile external environment characterized by shorter product life cycles, heightened local and global competition, maturing industries and flat markets, and a quickening pace of technological developments, many companies are in a position where NPD is no longer a strategic option but has become a strategic necessity.

According to managers and researchers, the benefit resulting from improving the early phases of New Product Development which typically precede the detailed design and development of a new product are likely to far exceed those that result from improvements aimed directly at the engineering design process.

Several critical success factors for these first stages of NPD have been identified in the literature. Detailed customer needs analysis is one of these critical factors. Capturing customer needs is crucial when producing high quality products. Quality Function Deployment (QFD) is a method for getting in touch with the customer and for using this knowledge to develop products which satisfy the customer.

Quality Function Deployment (QFD) was conceived in Japan in the late 1960s, during an era when Japanese industries broke away from their post-World War II mode of product development through imitation and copying and moved to product development based on originality. QFD came into being in this environment as a method or concept for new product development under the umbrella of Total Quality Control.

Today, QFD application goes beyond product and service design, although those activities comprise most applications of QFD. QFD has been extended to apply to any planning process where a team has decided to systematically prioritize possible responses to a given set of objectives.

I first came across QFD as a methodology in the mid-1990s. At that time, I had over 10 years' engineering experience and was a university professor of Industrial Marketing. The matrix approach which characterizes QFD seemed like a great way to keep track of the multitude of requirements and relationships that drive design decisions during the course of product development. I think that QFD is a powerful tool for delivering a valuable product to customers and leading to significant improvements in product/process performances. It is a team-based system which means that team members work closely enough with one another to provide accurate and useful appraisal information. This implies that not only the voice of the customer but also the voice of all firm's different departments are taken into consideration.

The book provides a clear description of a comprehensive quality function deployment framework. The reader is guided on how to create the matrices through practical examples. Discussions are also provided on how to gather information on customer needs and how such information could be used in product or service design and helping a firm gain ground compared to its competitors.

These issues are addressed by the author in a systematic way and with a distinctly analytical approach, but always using accurate and understandable language.

The presentation is also accompanied by a considerable amount of explanatory examples, taken directly from experience gained in the field.

In this sense, the book is an example of a perfect blend of technical expertise and professional experience. Consultants, practitioners, engineers, and students will find this book a useful reference manual and a good introduction to quality management and to quality function deployment in general.

Padova, Italy, July 2014 Roberto Panizzolo

Acknowledgments

I am particularly grateful to Springer and to Maria Cristina Acocella, Assistant Editor at Springer, for all her assistance and also as she was the first to have faith in this project. Her courtesy, willingness, and precision have been a great help to me. My grateful thanks also go to the team of professionals and colleagues who helped me in the last months of work on the book: to Mary Campbell, our meetings played an important role in revising the text; to Paola Scagnellato, my mother, whose great ability for language and reference research I have only recently discovered; to Mariangela Crotone, who dedicated a lot of her working time to keeping my progress up to schedule; and to Christiane Roethe, who helped me on one case study. Special thanks should also be given to Alessandro Trevisan, web survey specialist and ground source heat pump expert, for his steadfast support during these months. My thanks go to Danilo Tommasi who helped me develop the fastest QFD project ever. I would also like to extend my thanks to Roberto Panizzolo, with whom I shared several research and consultancy projects. I wish to thank all the students I taught and worked with, now professionals and engineers, for their contribution to QFD research and creativity.

Contents

Quality Function Deployment (QFD): Definitions, History and Models

1

Abstract

In this chapter the basic concepts of Quality Function Deployment are presented together with the context in which the method can be applied. In more detail, following a 5Ws 1H scheme, Where, What, Who, When, Why, How (5W1H), there are some basic questions to be answered:

- Where can QFD be applied?
- What is QFD? Who is involved?
- When was QFD invented? What is QFD history?
- How is it built? Are there any reference models?
- What are the strengths and weaknesses of Quality Function Deployment in today's research? Can QFD help creative, new product development?
- What is the QFD framework suggested in this book?

What you will learn in this chapter is primarily the initial context in which this method appears, the New Product Development (NPD) Process, some key definitions, the historical bases and two reference models known as "four phase" and "comprehensive" models. A summary of the main strengths and weaknesses of QFD is then presented. A section is dedicated to the capacity that QFD can have in supporting the development of "creative" products and services. And at the end there is a proposal for a practical framework with a good capacity to cover the different phases of the product development process and sufficiently easy to implement, even when there are limited resources to be allocated, and/or a short development time.

D. Maritan, *Practical Manual of Quality Function Deployment*,
DOI 10.1007/978-3-319-08521-0_1

1.1 Introduction

Understanding other people is a fundamental human need, answered by social interaction between human beings who, for this reason, in addition to body language, have developed the useful tools of speech and the written word over the centuries. The problem however is that languages differ both from each other and from country to country and the complexity of meaning varies according to a host of variables. It is therefore very difficult to fully understand the message that another person, from a different cultural background, wishes to convey. It is hard enough for each one of us to be fully aware of our own feelings or our deepest emotions.

On a daily basis we all find that conflict emerges from misunderstandings or the lack of communication between different groups of people, such as within a family or wider social groups, (e.g. between companies, customers, suppliers or between fans of different teams), each with their own cultural, linguistic or religious identities (think of the fear generated between cultures that do not know each other well), or between states (remember the two terrible World Wars last century).

Even if actually solving this problem is clearly impossible, there are however several ways to reduce it to levels deemed acceptable:

- Limit social interaction;
- Maintain relations only within a homogeneous group;
- Develop a common language;
- Activate mechanisms to allow for better understanding between the interlocutor and ourselves;
- Enable diplomatic strategies whereby there are no losers.

The first two points are simplistic and not applicable in the global village in which we live today, without becoming isolated. The others are currently used to create many effective relationships not only between different people, but also between people and the objects or services they make use of.

In his book 'Emotional Design', Norman (2004) explains that the principles required for the design of pleasant and effective interaction between human beings and products are exactly the same as those that underlie pleasant and effective interaction between individuals. In other words, the problems related to communication between people are also found in the relationship between people and products, as conceived, designed and produced by individuals, different to ourselves.

In the development of a new product there are methodological solutions, which allow for a better understanding of customer requirements and subsequently sound management of the technical specifications, quality, costs and engineering aspects. The theme is to simplify, make the information flow between different contexts that are often not homogeneous, both clear and understandable, i.e. between the customer and the supplier of the product. Here, the Japanese methodology Quality Function Deployment (QFD) comes into play.

This book is divided into five Chapters.

In this chapter we will discover the basic concepts of Quality Function Deployment (definitions, history) and the context in which the method can be applied (new product development process, strengths and weaknesses of QFD). One section is dedicated to the capacity that QFD has for supporting the development of "creative" products and services. At then end of this chapter there is a practical comprehensive QFD model.

In Chap. 2 the first two phases of the QFD framework will be discussed: Strategy (three matrices for targets, core competencies and customer segments) and Customer (*gemba* interviews, questionnaire survey, Preplan matrix).

In Chap. 3 the rest of the QFD framework phases will be discussed: Product characteristics, Functions and Mechanisms, Innovation, Parts, Costs and Production process, Reliability.

Chapter 4 deals with a simplified approach to fuzzy mathematics applied to QFD. The aim is to give the reader some basic tools to be used if the team wants to manage fuzziness in QFD projects.

Reading Quality Function Deployment case studies can be useful for understanding how professionals applied QFD and what the specific findings were. In Chap. 5 eight real QFD projects are presented, four in detail and four others briefly. Each case has its own peculiarities that will be discussed.

1.2 What You Will Learn in This Chapter

In this chapter I will present the basic concepts of Quality Function Deployment and the context in which the method can be applied. In more detail, following a Five Ws scheme, Where, What, Who, When, Why, How (5W1H), there are some basic questions to be answered:

- Where can QFD be applied?
- What is QFD? Who is involved?
- When was QFD invented? What is QFD history?
- How is it built? Are there any reference models?
- What are the strengths and weaknesses of Quality Function Deployment in today's research? Can QFD help creative, new product development?
- What is the QFD framework suggested in this book?

What you will learn in this chapter (Fig. 1.1) is primarily the initial context in which this method appears, the New Product Development (NPD) Process, some key definitions, the historical bases and two reference models known as "four phase" and "comprehensive" models. A summary of the main strengths and weaknesses of QFD is then presented.

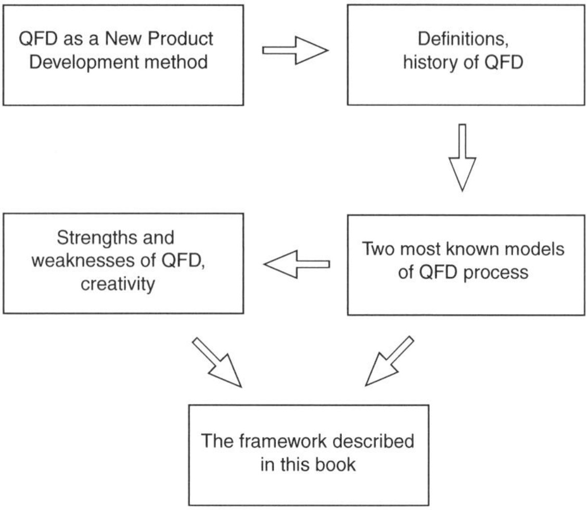

Fig. 1.1 Workflow of this chapter, QFD in NPD process, definitions and history, models, strengths and weaknesses, creativity and QFD, the framework described in this book

A section is dedicated to the capacity that QFD can have in supporting the development of "creative" products and services. And at the end there is a proposal for a practical framework with a good capacity to cover the different phases of the product development process and sufficiently easy to implement even when there are limited resources to be allocated, and/or a short development time.

Some Sub-chapters can be difficult to read because of the reference citations.

1.3 The New Product Development (NPD) Process

In the context of specific skills that are perceived as crucial today and in the future, a key role is performed by the ability to develop products and services at an ever increasing speed, higher quality at a reasonable cost and, ever more important, paying close attention to environmental and ethical issues. These latter, together with Responsible Business Practice (RBP) and Corporate Social Responsibility (CSR), play an increasingly significant role in public discourse on the governance of globalization, in particular as the transnational organization of production has accelerated over the last decades (Fransen 2013). The spread of Internet and the speed of information transmission, along with citizens' pressure group tools, such as the petition platforms change.org and avaaz.org, mean that companies (and people, managers, employees), the brand, and the entire new product development chain, are all coming under the watchful eye of ordinary citizens themselves, be they clients or not.

In this environment, successful development and commercialization of new products over time are essential to sustained competitive advantage of firms (Mu et al. 2009).

1.3.1 Products and Services

First we should clarify the term "product", and how it will be used hereafter, and to do that we need to understand what differentiates a physical product from a "service".

There are two main aspects:

- A service is intangible compared to a product. So, just as a product is an object identified by specific physical/geometric characteristics, a service is identified only by performance. It is difficult to standardize a service and therefore to apply the concept of conformity to specifications, that are typical of the manufacturing sector. Furthermore, it is complex to evaluate and manage the quality of a service: tests and checks are very complex, as complex as understanding how the end user perceives the quality offered. (Zeithaml 1981).
- Contemporaneous supply of a service contrasts with the sequentiality of production processes and consumption of the product. A service can not be separated from the context of its supply and consumption while a product can be. A service cannot be stored. As a consequence, lack of quality in a service is directly transferred to the consumer, without the possibility of correcting this (Lehtinen and Lehtinen 1982). The impact of human behaviour on a service is very strong, but less on a product. With a service a coherence problem exists concerning the behaviour of all the people involved (Booms and Bitner 1981) and their reliability (Franceschini 2002).

In spite of management difficulties inherent in a service, today we could agree with Clark and Fujimoto (1991): the customer consumes an experience delivered by a product rather than the physical product itself. There is a growing trend in manufacturing to move towards highly customized products, to be offered in a volatile environment. The concept of mass customization refers to a customer co-design process of products and services; buyers of manufactured products want more than the physical product itself (finance options, insurance, expertise to install, support, etc.) (Camarinha et al. 2013a). In other words, we, as customers, always want a complex blend of product and services, which can be more or less inclined towards a physical object or a service. In a service-enhanced products development the traditional functionality of a product is extended by incorporating business services for better response to customer needs and greater forms of product differentiation (Camarinha et al. 2013b).

For all the above reasons, from here on, by the term *product* I mean a complex mixture. I therefore advise the reader not confuse the use of the term "product" as used in this book with the more traditional meaning of its physical aspects.

1.3.2 Definition of the NPD Process

We could describe this process as the group of tasks and activities, methods and tools that are implemented when company wants to develop a product or a service, to be sold or supplied. The activities are phases that can be deployed in parallel, with partial or full overlapping or sequential, step by step. In the case of overlapping it is common to use the words 'Concurrent Engineering' (CE). Methods are systematic approaches used by development professionals. The physical tools can be computers, web systems (like survey web platforms), CAD-CAM framework, smart prototyping, 3D printers or others.

What phases form the New Product Development Process?

Scientific research has proposed several different models (see for example Mahajan and Wind 1992, Booz 1982, Clark and Fujimoto 1991).

According to the so called 'stage gate' framework (Cooper 1990), NPD could be divided into five stages: preliminary assessment, detailed investigation, development, testing and validation, and full production and market launch. In this approach, that was common in the Nineties, the output of the last stage is the input of the next stage; gates are quality control checkpoints; phases in every stage can be sequential or overlapped, in this second option they take place 'concurrently'.

Another interesting simplified depiction of the NPD process comes from Herstatt et al. (2004), who divide NPD into five phases:

- Phase 1 Idea generation and assessment;
- Phase 2 Concept development, planning;
- Phase 3 Development;
- Phase 4 Prototype development and testing;
- Phase 5 Production, market introduction and diffusion.

The product development starts with an idea, during phase 1 the idea is evaluated; after some iterative loops a rough assessment could be made according, for example, to company strategy, the market, the technology and the investments planned. In phase 2 a detailed product concept is deployed and the project is planned. In phase 3 the actual development of the product is done together with phase 4 when the prototypes are built and tested; in phase 5 pre-production is completed and the product launched in the market.

Finally, we can consider the structure proposed in Clark and Fujimoto's (1991) book Product Development Performance, focused on the American and Japanese automotive industry, where they consider product development as "a process by which an organization transforms data on market opportunities and technical possibilities into information assets for commercial production". The authors identify four phases of product development, intended as a process that can be symmetrical to the future customer's experience:

- Product concept. It anticipates the future customer satisfaction (rough technical targets and styling, forecast customer needs, preliminary process feasibility).
- Product plan. It specifies product function (major component choice and mechanical prototype, process feasibility, mock-up, styling, target costs, first technical trade offs).
- Product design. It represents product structure (detailed design, component design, detailed prototype, functional tests).
- Process design. It represents the production process (pilot production line, pilot vehicle).

From Clark and Fujimoto we learn that new product development processes can also be explained as a simulation of the link between product and customer. Now we shall try to analyse the variables that could affect this process and where Quality Function Deployment comes in.

1.3.3 Critical Variables in the NPD Process

As in every complex environment, the New Product Development process changes with different management approaches, tools and methods, which can vary from case to case, from company to company.

Now, let us consider the NPD model, presented by Muffatto and Panizzolo (1992); it identifies six critical variables that influence the NPD process:

1. Strategy;
2. Co-design and supplier involvement;
3. Organizational structure and technical staff;
4. Methods;
5. Design technologies;
6. Performance.

Let us examine these in detail.

(1) NPD Strategy. The product development strategy is, above all, related to the way product and service innovation is carried out (leader or follower innovation for example); but strategy also means how the company manages competitor analysis, market analysis and their relationship with technical design. For example, some researchers found that for highly innovative technical products, strategy development proficiency is related to the proficiency with which research, manufacturing and marketing activities are performed and to the degree of organizational integration (Millson and Wilemon 2010).

(2) NPD Co-design and supplier involvement. Co-design above all means early supplier involvement, i.e. collaborative innovation with suppliers during the first phases of the NPD process. What seems to be clear is that suppliers dominate component innovation, whereas buyers lead on architectural

innovation and that a second variable distinguishes between radical inputs and incremental inputs (Rosell and Lakemond 2012); in this way suppliers behaviour may vary in function of these two aspects.

On the other hand these development partnerships are not always harmonious and they are frequently fraught with miscommunications and crisis; the main problem is that most of the time managers lack guidance on how to recognize and handle crisis (Lynch et al. 2013).

(3) NPD organizational structure and technical staff. The way different projects are managed inside a firm, on the whole and individually, affects new product development. In the twenty-first century, Project Management (PM), developed in the twentieth century as a group of methods, tools and behaviours, has received some criticism over the last few years. In the past, PM skills, rules and activities evolved into various Bodies of Knowledge, the most widely adopted being the Project Management Institute's PMBOK, which tried to standardize practices and norms for new projects. These Bodies influenced the way projects have been planned. Today something is changing. For example, some authors, as a part of the Rethinking Project Management (RPM) study conducted by Sauer and Blaize (2009) about project management in the Information Technology industry, concentrated on the communication process inside and outside development teams for IT projects, that are characterized by complexity and uncertainty, are subject to change in turbulent business contexts, are more pervasive and integrated with other activities and are critical for the competitiveness of an enterprise (McKay et al. 2014). What it seems to be emerging is that the old rules are probably not valid for this rapid changing present. But new rules haven't been defined yet.

(4) NPD methods. Now we can summarize some of the numerous acronyms, which describe a variety of methods in some way involved in the New Product Development Process, to manage lead time, quality and costs. There are a great deal of them, many come from Japan and have been assimilated and adapted in the West, others are considered traditional. Gantt Charts (GANTT), Critical Path Method (CPM) and Program Evaluation and Review Technique (PERT) are adopted in project management activities to optimize project phases and find the time critical ones. Methods like Quality Function Deployment (QFD) and Analytic Hierarchy Process (AHP) can usually be adopted during marketing, market analysis and product planning activities to manage quality perceived by customer. Phase Review (PR) or Design Review (DR) can be used (as gates in the stage-gate approach) to check the work of the activity and create memory of errors for future similar projects. Design Review Based on Failure Modes (DRBFM) is an approach developed in the automotive industry (see www.aiag.org). Failure Mode Effect Analysis (FMEA) is a preventative approach used for products and processes in order to organise and manage potential failure modes, considering the effects, the probability of occurrence, the seriousness of the failure and the likelihood that the defect will reach the customer, the risk priority measure and finally the recommended corrective actions and an improved risk priority ranking. Advanced product quality

planning (APQP) is a framework of procedures and techniques, particularly adopted in the automotive industry, that involves adoption of FMEA, Statistical Process Control (SPC—Robust Design), Measurement Systems Analysis (MSA) and Production Part Approval Process (PPAP) methods (AIAG 2012). Value Engineering (VE), Value Analysis (VA), Design to Cost (DTC), Variety Reduction Program (VRP) are some of the methodologies, used to control costs during new product development.

And what about the method called Design for Six Sigma (DFSS)?

DFSS could be seen as a group of tools like QFD, FMEA and Robust Design modelled into an improvement process that would assure both customer satisfaction and competitiveness, preventing problems instead of just fixing them (Mazur 2002). On the other hand, some researchers such as Watson and De-Yong (2010) found problems in the alignment with the engineering design process for all the DFSS models investigated, suggesting that, so far, a coherent body of knowledge about DFSS is not clear and that it should be developed for the future.

(5) NPD design technologies. Let's consider the technological support to engineering and more in general to product development. The digital age we are currently living in offers potentials we could not have foreseen just a few years ago, and development in I.T. is accelerating, bringing with it both new opportunities and risks in every human sector. If we refer to a NPD process, in addition to integrated CAD—CAM systems, which are increasingly more powerful and accessible, 3D printing gives us the possibility to speed up the prototype phase, changing the virtual renderings of few years ago into real, tangible objects. Internet allows people living in different locations to work together in spite of distances.

In the next chapter we will see how, by using internet, it is now possible to carry out market research surveys at very affordable costs and with amazingly quick results. In methods like QFD, open source software allows for calculations to be made using quite a simple program and therefore eliminating the need to buy a dedicated program.

(6) NPD performance. There have been several attempts to measure the performance of product development. Since the 1950s, researchers have been trying to measure the NPD process to define its "best practices" (Carter and Williams 1957). Since SAPPHO project in the seventies (Rothwell et al. 1974) there has been a continuous and systematic comparison of successful innovations to unsuccessful ones in the same market. (Gerstenfeld 1976; Rubenstein et al. 1976; Rothwell 1977; Cooper 1979).

During the eighties and nineties, the dominating subject was the decline of Western economies compared to steady growth of Japanese products in world markets. This created a growing interest in the Japanese approach to quality, NPD management, tools and lean production (Pascale and Athos 1981; Shonberger 1982). At the end of the twentieth century, market globalisation and production de-localisation to new developing markets brought more changes in ways of developing products. Today, contemporary organizations, faced with global competition and external environmental turbulence, require highly creative NPD teams to survive (Fain and Kline 2013), more focus on technology and technical innovations, to gain and maintain competitive strategic advantage (Haverila 2011).

But product innovation is an activity with high investment and risk, and a low success rate.

How to increase the chance of new product success?

As a result of their research, Cooper and Kleinschmidt (1993) suggested that "the greatest differences between winners and losers were found in the quality of execution of pre-development activities". These activities are also known as "Fuzzy Front End" (FFE) or, in other words, "up-front activities", "pre-development phase", "pre-project activities", "concept development"; the word "fuzzy" describes the fuzziness (uncertainty and ambiguity) in new product development (Zhai et al. 2012). According to the Herstatt et al. (2004) model presented before, FFE activities involve phase 1 and 2, idea generation and assessment, concept development, planning; output of the fuzzy front end is a detailed business plan, which is the basis for the decision on a business case (Herstatt et al. 2004).

Cooper and Kleinschmidt (1993) believe that *pre-development activities* (Product Concept phase in Clark and Fujimoto 1991), especially those related to marketing, have great influence on the time, cost and quality of product development. Cao et al. (2008) reported on key variables for success, with regard to the FFE activities of NPD in 513 Japanese companies, and they confirmed the hypothesis of Cooper and Kleinschmidt: "the more both market uncertainty and technical uncertainties are reduced during front end, the higher is the effectiveness of NPD projects" and "the more intensively new projects are planned prior to the start of development, the more both market and technical uncertainties are reduced".

A study by Cao and Zhao (2009), focused on 160 NPD projects from Chinese manufacturing firms, and was aligned with previous research. These authors recognized that Western (US and European) researchers and managers of innovation have also paid attention to FFE activities; they identified the front end as "being the greatest weakness in product innovation".

We learn that today, Fuzzy Front End activities seem to be a strong area for a successful NPD process. And later on we'll see that Quality Function Deployment can give good support to FFE activities.

Finally, we might ask ourselves if QFD can be used in environments outside NPD. The Quality Function Deployment method came into being in order to develop new products but it can be extended to a wide variety of neighbouring situations, like quality management, customer needs analysis, planning (strategy, business, research, process, supply chain), decision support systems, management support, workgroup promotion (Chan and Wu 2002).

1.4 Quality Function Deployment Definitions

Overall, we could describe Quality Function Deployment (QFD) as a powerful, proactive decision support method applied in relationship-wise contexts, probably the most powerful. It has been developed with the economic environment in mind and it is based on a sequence of bi-dimensional mathematical matrices, which link together some NPD areas, often different ones. The purpose is to calculate ranking numerical indicators, to be represented graphically and to create an easy to understand database, useful for the decision maker.

QFD is a "living method and [a] way of thinking…"; "a powerful method for creating a system of cooperation among… divisions" like product planning, design, manufacturing, sales (Mizuno and Akao 1994).

What does Quality Function Deployment mean? First of all, Quality Function Deployment (QFD) is the English translation of six Japanese ideograms which were gradually introduced during the 70s by Yoji Akao and Shigeru Mizuno (who coined the term *kino*):

Hinshitsu (product attributes, quality or qualities)
Kino (functions, mechanisms, opportunity)
Tenkay (deployment, development)

Several researchers have given their own definitions of QFD. Let's look at some of them.

If we consider the environment where QFD can be applied, apart from new product development, this method needs to be linked to a Total Quality Management (TQM) process such as Company Wide Quality Control (CWQC), where customer satisfaction is placed as principle number one (Ginn et al. 2010). This philosophy features a sharp focus on the customer, cross functional management and the process rather than product orientation; from this point of view, QFD is "a method of continuous product improvement, emphasising the impact of organizational learning on innovation" and it "becomes a management tool to model the dynamics of the design process" (Govers 2001).

Akao (1990) states that QFD "provides specific methods for ensuring quality throughout each stage of the product development" with an emphasis on the design stage. It is a customer oriented method that translates the requests of customers into design and quality targets. Shigeru Mizuno focuses his attention on the term "functions": the deployment of quality functions is a step-by-step path that systematically deploys the functions of the company's process and the activities that contribute to quality and that utilize objective rather than subjective procedures (Akao and Mazur 2003).

The American Supplier Institute (ASI 1987) considers Quality Function Deployment as "a system for translating consumer or user requirements into appropriate company requirements at every stage from research, through product design and development, to manufacture, distribution, installation and marketing, sales and service" (Xie et al. 2003).

We could say that QFD represents the deployment of the attributes of a product or service desired by the customer through the functional components of an organisation: it is a product/quality/service/process/software planning process (ReVelle et al. 1998).

But QFD is also a method of focusing company organisation on those product characteristics that are more important for the customer (Bellenda 1992).

To summarise, Quality Function Deployment:

- has excellence of product/service quality as its main target (in an extended sense of the word "quality"), assessed as the level of customer satisfaction reached;
- uses objective mathematical methods or, at least, as objective as possible (as we will see later, in the QFD working team evaluations, a subjective component is present anyway);
- can be deployed alongside the entire product development process, from market and customer analysis, to design, process engineering and pre-production (nevertheless we will see that this method can be interrupted and/or modified, without jeopardising results obtained up until that time);
- tries, also and above all, to find a correlation between non homogeneous contexts, for example customer's needs and product characteristics, specified during design stage;
- aims to involve each company function in the design process;
- helps management and work groups to come to a decision, that can be put into action, and to make any necessary compromises; it gives a clear indication, a solid, complete pathway to follow;
- finally breaks down complex and foggy problems into an orderly structure of small ones, much simpler to solve.

1.5 QFD History

QFD was developed in Japan, and came into being at the end of the sixties (Akao 1990) as a new method or concept to develop products, amongst the various activities of Total Quality Control TQM (Akao and Mazur 2003). There were two main reasons for developing this method: (1) how to project a product which can meet customer requirements and (2) the need to draw up quality control charts (QC process chart) of the process, before actual production (Cristiano et al. 2001). During the fifties in the United States and in other countries, matrices were used for a wide variety of purposes (ReVelle et al. 1998), but it was in Japan at the beginning of the seventies, that Nishimura (1972) presented an article about using a particular chart in the Mitsubishi Heavy Industry plant in Kobe, which was applied during the design of large cargo vessels, called Quality Chart: the main purpose of this was to improve the quality of the product. Shigeru Mizuno, Yasushi Furukawa and Norikazu Mizuno indicated a method of development for this company which, afterwards, Yoji Akao called "quality deployment" (Mizuno and Akao 1994). At the same time, in April 1972, Akao, introduced a concept summarised by four Japanese ideograms, *hinshitsu tenkai*, where the first guidelines of the method were explained (ReVelle et al. 1998). QFD became a key procedure for Toyota, starting from Toyota Auto Body (Akao and Mazur 2003), initially using it to reduce and control the huge problem of car body corrosion, badly afflicting Japanese cars in the sixties and seventies, as costs for repairs and warranties were exceeding a factor of four on company profit (Xie et al. 2003). Akao reported on QFD applications during the seventies in the following companies: Dynic, Hyno Motor, Sanyo Seikio, Ricoh, Pentel, Daihatsu (Mizuno and Akao 1994). In 1978 Mizuno and Akao published their first book about QFD, translated into English in 1994 in which Mizuno talks about the functions of quality.

Should QFD remain within the confines of Japanese production?

Different authors (Xie et al. 2003; Mizuno and Akao 1994; Akao and Mazur 2003) recognise 1983 as the year of the introduction of QFD to the western world (USA and Europe), at two events. Akao, Kogure Masao and Yasushi Furukawa were present at a seminar held in Chicago attended by about eighty people, most of them from big American companies, some others from Sweden and Italy (Akao et al. 1983). *Hinshitsu tenkai* was initially translated in 1972 as "quality evolution" at the suggestion of L.T. Fan, a researcher at Kansas State University, but during the seminar it was changed by chairman Masaaki Imai into "quality deployment" (Akao and Mazur 2003). After more than 10 years of studies and development in Japan, during October 1983 Kogure Masao and Akao published an article entitled "Quality Function Deployment and CQWC in Japan" in the magazine Quality Progress (Chan and Wu 2002).

In 1984 Don Clausing introduced QFD into Ford, where it is still implemented as a structural method (Ginn et al. 2010). Three new names appeared during this period: Bob King, founder of GOAL/QPC (Growth Opportunity Alliance of Lawrence/Quality Productivity Center: http://www.goalqpc. Com), John Hauser and Larry Sullivan (founder of the American Supplier Institute http://www.amsup.com), recognised among the first researchers, who contributed to the growth of QFD in the USA (Martins and Aspinwall 2001).

Since then there has been a steady increase of publications and applications, for example Xie et al. (2003), mentioning Sullivan (1986), tell us that during the same year, a survey conducted by the Japanese Scientist Union and Engineers, proved that 54 % of the member companies were using QFD, with special focus on high technology and transportation. Bob King writes, in the introduction to J. Terninko's book "Step by Step QFD" (Terninko 1997): "from 1975 to 1995, this tool/process was integrated with other improvement tools to generate a mosaic of opportunities for product developers".

QFD therefore is associated with other methods, such as statistical Taguchi methods for reduction of "noise" or the AHP Analytic Hierarchy Process (we will talk about the method in the next chapters) or FMEA analysis, Failure Mode Effect Analysis, in order to give integrated support to project planning.

The term "House of Quality" (HoQ), refers to a particular matrix or quality chart, closed at the top with a triangular "roof", and came into being in Toyota Auto Body thanks to the imagination of Tsuneo Sawada, who introduced it in 1979 to the Japanese Society for Quality Control JSQC conference (Akao and Mazur 2003). The House of Quality was to become, for the western world, the symbol of QFD. In 1988 Hauser and Clausing wrote an article entitled "The House of Quality" for the Harvard Business Review, which became famous and contributed to QFD popularity. A four matrix process was presented, which was often used in cases illustrated in scientific literature, and that will be discussed later.

QFD is rich in matrices, as we will discover in the next chapters. They are adaptable and can be modified according to specific cases. It is both a pity and a limitation that many QFD users just concentrated on one single matrix, even though it is the most famous one.

In 1990 Akao, published an important text about the QFD method, a book entitled "Quality Function Deployment QFD, Integrating Customer Requirements into Product Design" translated into English by Glenn Mazur (Akao 1990). It presents a framework, a very complex path made up of matrices, named *Comprehensive* QFD. Starting from the customer, it analyses all the technical aspects (features, functions, mechanisms, parts) costs and reliability, finally linking these with production quality control. It is the inspiration for the development of the framework suggested in this book.

Glenn Mazur's work has allowed for the spread of QFD, still going on nowadays. In 1993 the Michigan QFD Institute, Michigan, (USA) was founded, with the official approval of Akao.

From the mid nineties to the early years of the new Millennium some authors reported that more than 1,000 case studies were published all over the world (Akao and Mazur 2003).

During the first decade of the new millennium some reviews of scientific literature on QFD were presented.

I recommend them as interesting reading to understand how lively the discussion on the subject is. There are also the first comparisons on how QFD is implemented in different contexts, such as the eastern or the western world, and the results obtained. In 2002 Chan and Wu presented a study that analyses about 650 publications, gives internet references and tries to organize the subject under discussion into macro–areas. Apart from the excellent work of the Hong Kong authors, I can also cite the work of Cristiano et al. (2001) on a group of 400 firms about different approaches, both Japanese and American, to QFD, with the surprising discovery that a higher percentage of American companies use QFD compared to Japanese ones, probably thanks to the great multi-functional integration involved in the use of QFD. QFD becomes a powerful tool for organisational change, improving cooperation and internal dialogue. Carnevalli and Miguel (2008) recently wrote a very useful review. They are Brazilian researchers who discussed difficulties caused by calculation matrix and implementation of the method, identifying 157 articles and making an analysis of publications from 2000 to 2006.

Asia, geographically close to Japan, has been an area of development of QFD starting from the nineties, with applications in Taiwan, China and India (the latter for software development and the car industry (Akao and Mazur 2003).

And Europe? In Europe Italy seems to be the first country where QFD is being implemented, following some lectures given by Akao at Galgano and Associated in 1987. However publications and case studies available up to today are still very limited compared to other countries.

Govers (1996) reports a few applications in Holland, United Kingdom, Sweden, Austria, Spain and most of all in Turkey and Germany, the latter thanks to the activity of Koln University (Akao and Mazur 2003).

Let's come to the present.

Scientific activities of the last decade seem to be concentrating mostly on the improvement of vagueness, which is imprecision, of potential customer evaluation and of the choices that the QFD team put inside the matrix, the so called fuzzy mathematics, an area in which eastern research seems to lead. Apart from the fuzzy theme one should highlight the research being carried out with the purpose of improving (and simplifying) the QFD method, finding its limits and suggesting possible solutions. Verification through a scientific "renting" website deepdyve.com has shown how more than 100 articles on QFD were published in 2013, demonstrating that there is still great interest in this method of product planning. Lastly, the extreme development of technologies and web-based services force users to adapt the method to the quick developing times of software. The challenge of the digital era in twenty-first century sees QFD as a key method in order to build quality in these I.T. systems and understand future requirements (Akao and Mazur 2003).

1.6 Four Phase and Comprehensive Models

Practically Quality Function Deployment is made up of:

- bi-dimensional mathematical matrices, also called "charts", that correlate two different NPD environments or the same environment (we will learn how to use the Analytic Hierarchy Process methodology, AHP);
- a model, in other words a framework, that connects the matrices.

Two models are the best known:

- the four-matrix framework, proposed by Hauser and Clausing (1988) and
- the Comprehensive framework, proposed by Akao (1990), with a lot more matrices that build the "Matrix of matrices".

The first model, simplified from Akao's original one, seems to be described in more detail in scientific literature; the second is considered gigantic and nearly unattainable by some authors (Xie et al. 2003; Cohen 1995).

First of all let us analyse the four matrix model.

1.6.1 Four Phase Model

As we can see in Fig. 1.2, four matrices are used sequentially, deployed to cover the new product development process, from customer demand to production (Hauser and Clausing 1988); as you can imagine, the output from each previous matrix constitutes the input for the following one.

The first matrix, also called "*House of Quality*" (HoQ, Fig. 1.3) is the best known and it is considered a central element in the QFD method (Xie et al. 2003). In fact, according to researchers Cristiano et al. (2001), in the new millennium most of the US companies that have adopted QFD, only use the House of Quality matrix, unlike Japanese companies; these last, in general, work with an extended group of matrices and with a holistic vision of the internal processes. In a recent review of published papers on QFD, about 33 % of the articles considered concentrate on HoQ (Carnevalli and Miguel 2008).

Although we will see the structure of this kind of matrix in detail in the next chapters I think that the House of Quality should to be presented here in a simplified way.

House of Quality is a matrix that correlates Customer Requests, also called "Demanded Qualities" (DQ) or "Whats" or "Customer attributes" (CAs, Hauser and Clausing 1988) or "Customer Requests" positioned on the rows, to the technical characteristics, i.e. the performances of the product, also called "Quality Characteristics" (QC) or "Hows" or "Engineering characteristics" (ECs, Hauser and Clausing 1988), placed on the columns. The central body of the matrix has a certain number of cells, the number of rows multiplied by the number of columns, where some symbols can be inserted, to symbolize the link between Whats (customer's Demanded Qualities) and Hows (product's Quality Characteristics): the symbols can express absence of correlation (no symbol), weak, medium, or high correlation.

The matrix roof is made up of cells that contain the positive or negative link between each characteristic, compared to all the others. Effectively, implementing a characteristic can advance or hold back the performance of another characteristic. For example, in a car you can think of the links between the characteristics "weight", "fuel consumption" and "safety"; weight and fuel consumption are negatively correlated, if weight increases, performance in consumption decreases (remember that the direction of improvement of fuel consumption is negative); on the other hand, weight and safety are positively correlated.

The House of Quality foundations are built on the importance of the characteristics (Hows-QC), calculated considering their ability to meet the customer's

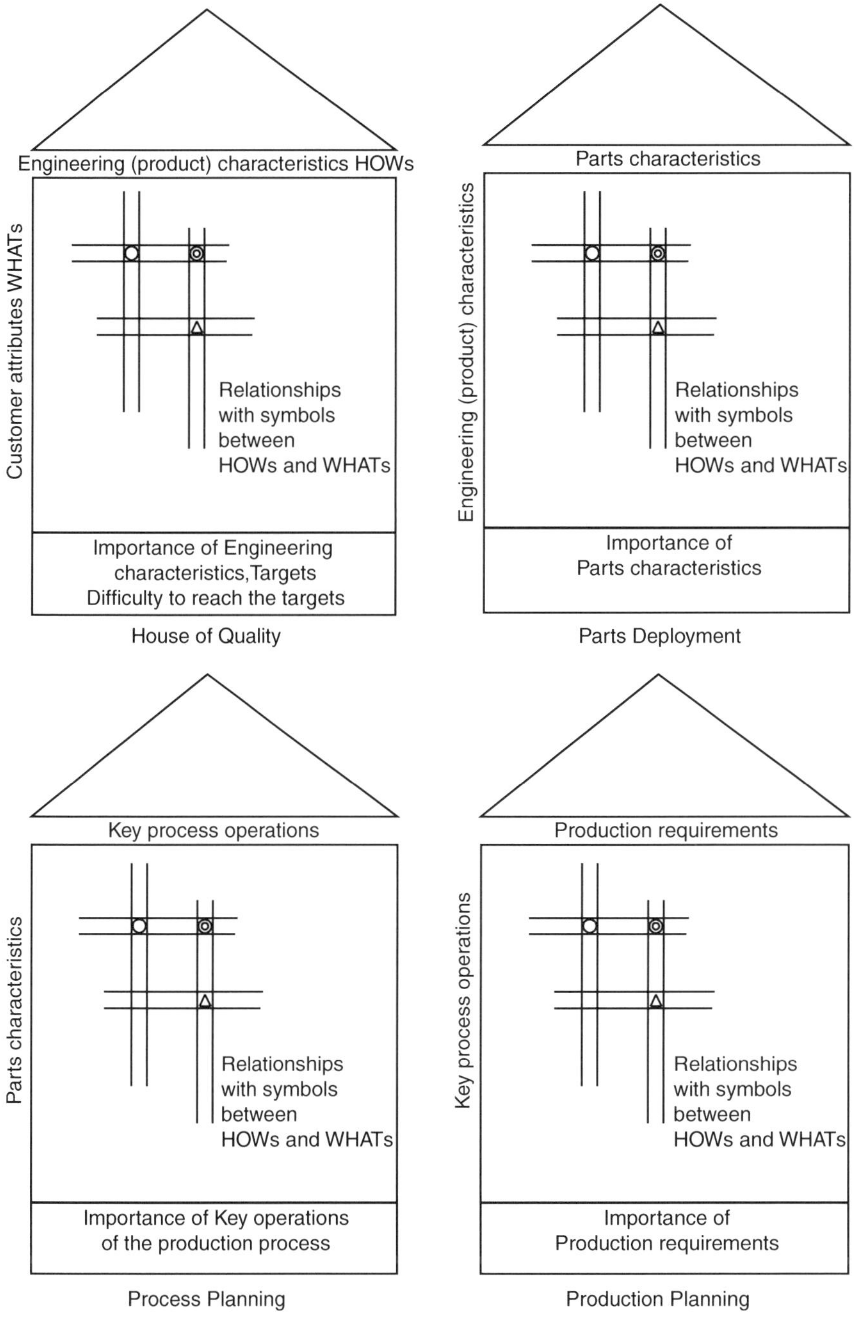
Engineering (product) characteristics HOWs
Customer attributes WHATs
Relationships with symbols between HOWs and WHATs
Importance of Engineering characteristics, Targets Difficulty to reach the targets
House of Quality
Parts characteristics
Engineering (product) characteristics
Relationships with symbols between HOWs and WHATs
Importance of Parts characteristics
Parts Deployment
Key process operations
Parts characteristics
Relationships with symbols between HOWs and WHATs
Importance of Key operations of the production process
Process Planning
Production requirements
Key process operations
Relationships with symbols between HOWs and WHATs
Importance of Production requirements
Production Planning

Fig. 1.2 Four phase QFD model (adapted from Hauser and Clausing 1988)

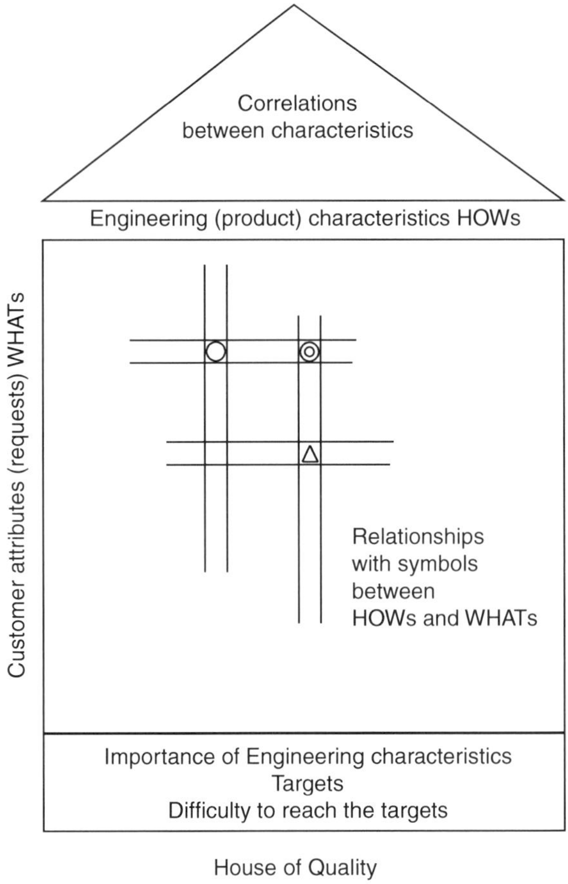

Fig. 1.3 House of Quality, Customer Request (rows)—Characteristic (columns) matrix

requests (Whats-DQ); furthermore, the technical targets of these characteristics as well as the difficulty in reaching them are present here.

Who fills in the House of Quality, evaluating what symbols to be inserted? Ideally a cross-functional team (Xie et al. 2003), made up of professionals that come from different offices: sales, marketing, design, engineering, quality assurance, production.

Now we can move to the second matrix. You can see that the Hows (the columns of the HoQ) become the Whats (the rows of the second one); the new Hows are the Characteristics of the Parts of the product (not the parts themselves but the performance of each part). For example, in a car a part characteristic could be the "thickness of the weather stripping". For the rest the structure and the calculations are the same as in the HoQ.

A third matrix and a fourth one follow. The third one is focused on key product process operations, the Hows, such as for example, the "rpm of the extruder producing the weather stripping". The fourth matrix deploys production requirements in columns; here the "rpm of the extruder" is linked to the control setting, operator training, and maintenance (Hauser and Clausing 1988).

Although this process deploys four charts, in practice you can use the number of levels you want or whatever seems suitable (King 1989). We can see in this flowchart that, little by little, the principles that govern House of Quality are extended to every effort made to establish clear relationships between customer satisfaction and product production.

1.6.2 Comprehensive Model

Certainly Akao's model is more complicated than the Hauser and Clausing one. According to Akao (1990), a QFD system must reflect considerations about technology, reliability and cost.

In the so-called Matrix of matrices, four columns are present, showing:

- "quality deployment";
- "technology deployment";
- "cost deployment" and finally;
- "reliability deployment".

In 1999, Glenn Mazur (the translator of Akao's book) published some manuals, focused on product and service, which contributed to a better understanding of the comprehensive model. As a philosophy and approach, the Comprehensive QFD framework suits the Japanese business world better than the American or European ones. In 2001, Cristiano et al. described the two different approaches to the use of QFD. American companies are focused on House of Quality and new customers' requests rather than on monitoring the existing ones, or on improvement of the company's organizational functions or decision making processes. According to these same authors, Japanese firms use comprehensive models with several matrices and study the existing customers' requests (for example the warranty management). In this case the paradigm seems to be the extended model, which aims to improve, not only the quality of the product, but also the quality of the design process (Cristiano et al. 2001).

1.7 Strengths and Weaknesses of Quality Function Deployment

As with all solutions used by human beings to understand and manage natural phenomena or human interaction, every methodology also has vast areas for improvement, as well as limitations. The same consideration is valid for QFD. Let's discover what the strong and weak points of QFD are, using the results of recent research.

1.7.1 Strengths

QFD allows for maximisation of the level of Customer satisfaction (Poel 2007), in wider terms the interlocutor—the point of origin—giving input of the initial information (their expectations, their requests). While reading this book you will understand that QFD develops a common language, together with graphic and mathematical languages, it simplifies interaction both within an organisation and towards external interlocutors, it activates win-win conditions.

QFD can be a possible solution to the problem I mentioned in the introduction.

Several books and studies have presented a list of advantages either partially or totally induced by the use of this method.

To list them, QFD:

- helps in reaching compromises between customer requirements and what the company (organisation) is able to produce; it can also help in finding new opportunities and in defining strategies to increase profits and market share (Chien and Su 2003);
- identifies the check points for *gemba*, a Japanese word that refers to the place where source information can be learned (Akao and Mazur 2003); you will learn much more about *gemba* in the next chapter;
- facilitates or, rather, requires a strong inter-functional involvement between different offices, for example the sales department, technical department, engineering department, quality control and production department;
- allows for a reduction in start up problems;
- definitely enhances analysis concerning market, customers and competitors;
- clarifies check points, determining a reduction in development time (lead time) and better project planning and timing;
- through its matrix calculation flow it can fix targets from design to production; it can be stated that quality of the process is built "upstream" (Govers 1996).

Recent research (Carnevalli and Miguel 2008) has identified 235 benefits, reported in about 160 scientific publications, divided into: 20 % tangible (reliability, reduction in project changes, reduction of time for development, higher profit, reduction of customer complaints) and 80 % intangible (flexible tool, relationship between requirements and features, enhancement of communication, help for decision making, creation of an inter-functional team, increase and conservation of the company knowledge, increase of customer satisfaction).

We can summarise the impact of QFD in the business environment, using the words of Akao and Mazur (2003), themselves quoting Yoshizawa (1997):

> QFD has redefined quality control in manufacturing by moving it upstream to quality control for development and design. It has shifted the focus of TQM from process-oriented QA [quality assurance] to design-oriented QA and in the creation of product development systems. QFD has provided a communication tool for designers. […] QFD is a powerful tool for engineers to build a system for product development.

It is clear that QFD does not solve all technical and management problems, nor foresees the unforeseen; it does not bring with it the absolute truth about the future evolution of a relationship, of a product, or of a service. The subjective component, we humans, still remains strong. Below I will summarise the limitations of the method and the attempts we can make to overcome them, partially if not totally. QFD is in general, as stated, a good, schematic and quite intuitive informative support, in all the key phases where a decision from a person in charge or from a team is required.

1.7.2 Weaknesses

There are some difficulties in applying the methodology; the most evident are listed below. It is useful to make reference to some recent reviews, such as Carnevalli and Miguel (2008, 157 analysed articles) and Cristiano et al. (2001, panel of 400 firms), as well as Xie, Tan and Goh's excellent book, published in 2003.

A list can be drawn up and integrated with direct experience.

- There is an actual lack of knowledge on how to use the method practically (Carnevalli and Miguel, 2008, Martins and Aspinwall 2001). It would seem that there are few studies about improvement of training for using this method, or for the management of specialised areas, like technical bottlenecks or cost related aspects.
- Interpretation of market data seems to be difficult. How should one identify and organise what I will define in the next chapter as Voice of the Customer (VoC, Voce of the Customer), i.e. the combination of open answers from customers or potential customers, given during unstructured interviews? How should one consider the development of innovative products, integrating the VoC of the future, and not only the current one (Xie et al. 2003), that is imagining what customers will require in the future, even though they have not expressed their requirements yet? How can we give a value to market data (Carnevalli and Miguel 2008)? One of the problems linked to the dynamic nature of customer requirements (Cristiano et al. 2001) is sometimes solved using support methods such as Taguchi's (ReVelle et al. 1998) or the so-called "fuzzy mathematics". This latter method, together with the AHP method, seems to be the most promising research trend of the last few years, even if it is still to be verified whether these two approaches make QFD even more complicated (Carnevalli and Miguel 2008). Fuzzy mathematics and AHP will be discussed in the chapters that follow.
- A substantial lack of "longitudinal" studies has been highlighted, that is long term analysis of QFD application in the same firm. Carnevalli and Miguel (2008) considered this aspect very important, in order to study how long-term application of the method can influence performance and results. In Chap. 5 we will

examine a "longitudinal" case of QFD strategic analysis in a company dealing with renewable energy.

- A discrepancy seems to exist between, on the one hand how things work in companies, and on the other how things work in research; this is also confirmed by the two Brazilian authors (Carnevalli and Miguel 2008).
- Only 13 % of cases analysed by Carnevalli and Miguel extend beyond House of Quality and try to define procedures to adopt during project development and production. In their conclusion they state: "in spite of this flexibility, most of the studies limit QFD use to the development of the quality matrix (House of Quality – note of the author), which could restrict its results". It is, therefore, very important to understand how the method can be easily applied apart from just using the famous matrix. This is, in fact, one of the main purposes of the book you are reading.
- There might also be some organisational difficulties. QFD allows for different levels of performance and different results, according to both the organisational context and the new product development process to which it is applied, and to the strategic planning for improvement in medium and long-term (Griffin 1992). The person who has the greatest decisional power plays a fundamental role in this field. What I discovered during some of the applications discussed later in Chap. 5 is that this aspect is very close to reality. I can venture a guess by stating that QFD potentially performs better in very small, small and medium sized firms, because it is much easier for owners or top management to either totally embrace or refuse the application of the method; while in big companies decision processes are longer and more complex which can de-motivate the inter-functional working team or spread the QFD project over too long a period. For instance in their research published in 2001 Cristiano et al. show that in 100 % of the cases where prerequisites were defined, need for support from upper management was cited. It is also possible to assume that in addition to the ability to make investment decisions, there is also the factor of the size of the costs of the QFD project that need to be faced. Again, in Chap. 5, we will see that sometimes the QFD method can be implemented with very good results even with a rather small amount of money.
- Managing large dimension matrices is very complicated, even when algorithms have been used for the optimisation of the number or lines and columns. Carnevalli and Miguel make reference to Lowe and Ridgway's (2000) suggestion of limiting the dimension of the matrix to a maximum of 8 lines and 8 columns, even if it raises some doubts about the capacity to cover all the customer requirements and technical features of very complex projects. When working on some of our own QFD projects we dealt with very large matrices and can therefore confirm how our work was more difficult because of it.
- Some authors point out the problems that exist in the priority definition phase, i.e. the results from the calculations of the different matrices, such as the random setting of the number scale, conversion of the scale from ordinal to cardinal, and the lack of HoQ matrix roof utilisation (Xie et al. 2003). These latter authors raise doubts about the possibility of translating individual requests from customers into

collective requests, and about how to link the requests with specific features of the product, without breaking one or more logical calculation conditions (Poel 2007). Firstly, some more complicated than usual algorithms ones were proposed, based on a weighted sum of values: multi-attribute analysis, fuzzy management of the assigned weights in the HoQ matrix, and approaches which add weights and technical targets together (Xie et al. 2003). Secondly it is important to remember that customer requests are "product dependent", somehow they are linked to the product, for example, the request for "lower fuel consumption" in a car (Poel 2007); not only are they product dependent but they can also be conflicting.

Chapter 2 deals with some of the tricks useful in helping to find a solution to these observations, where the fundamental assumption is to understand that, conceptually, *Customer and Product are distinct entities*; the *semantic process of abstraction* (construction of a representative sentence) leads to needs, wishes and feelings that are personal to each customer, to which ideas from the working team are added, which have nothing at all to do with the product; in this way compromises between conflictual technical sentences are not needed, simply because in talking about Customer we do not talk about technical features.

The issues that are sometimes referred to as Customer Requests are, actually, still Product Features.

That is how "low fuel consumption" becomes "the desire to spend less money on journeys". In House of Quality, if we do not clearly separate the two contexts we run the risk of ending up not having a mathematically robust analysis.

On the previous pages I listed the weaknesses and strengths of Quality Function Deployment. To quote Griffin (1992) from his famous paper 'Evaluating QFD's Use in US Firms as a Process for Developing Products' in which 35 NPD-QFD projects were studied, the author says that QFD "is not a short-term panacea". However the following chapters in this book serve to aid the reader: in fact they have been written with these open points in mind, in an attempt to find some common sense, practical answers to these questions, and with the aim of excluding overly complex mathematical processes which would risk remaining only theory.

As far back as 1988 Hauser and Clausing stated at the end of their article, which presented the four phase QFD model (1998):

> None of this is simple. An elegant idea ultimately decays into process, and processes will be confounding as long as human beings are involved. But that is no excuse to hold back. If a technique like house of quality can help break down functional barriers and encourage teamwork, serious efforts to implement it will be many times rewarded.

1.8 Creativity and QFD

Is Quality Function Deployment a creative method? This is an interesting question.

If we look at the high level of use of mathematics and detailed analyses, the answer could be negative. But in reality things are very different.

In his book, Emotional Design, Norman (2004) identifies three types of design (interpreted as styling and development): visceral, behavioural and reflective.

- Visceral design concerns the aesthetic aspect of an object, for example its appearance, its shape, colours, characteristics that inspire our deepest and intimate emotions.
- Behavioural design refers to how effective the product or service is when it is used.
- Reflective design embodies what Norman calls the "rationalization and intellectualization of a product"; for example, reflective design, invites us to start a dialogue and discuss the object.

Certainly, design, a mixture of these three components, is connected with our emotions, including the feeling of beauty that the object generates, as well as the human cognitive process, which demands a certain utility of it and requires that it be able to efficiently perform the range of functions for which it was designed.

Emotions are linked to strong reactions to some external stimuli, for example, the need to survive or react to imminent danger. On the other hand the cognitive process is an attempt to understand the world that is around us. These two aspects, in and of themselves opposing, are strongly linked together, by the simple fact that they are part and parcel of our identity as human beings. In this sense an object probably acquires value because of its ability to start positive interaction (subjectively considered positive) and to generate self-fulfilment.

In reading Emotional Design the direct connection between QFD and the cognitive theme and behavioural design was natural. How then do we deal with our tendency to find emotional reactions to every object, whether animate or inanimate?

Considering the different phases of QFD methodology, and my role as teacher and consultant, I must admit that these things are more complicated than they seemed initially.

Referring to his visit to Swatch headquarters in Switzerland, Norman says:

> I learned that products can be more than the sum of the functions they perform: their real value can be in fulfilling people's emotional needs, and one of the most important needs of all is to establish one's self-image and one's place in the world.

We have seen that QFD can be considered as a method, that systematically analyses customer's requests and that translates these requests into characteristics of a product or a service. The process of defining customer requests, illustrated in the next chapter, induces us to mentally distance ourselves from the physical properties of the product and to create descriptive sentences that evoke feelings, personal needs, customer wishes, quite remote from the characteristics and functions of a product. The purpose of this distancing process is to correctly identify phrases (just a few words in each phrase, which we will call "Demanded Qualities"), that express the moods and feelings of our interlocutor.

I call this process "*abstraction*".

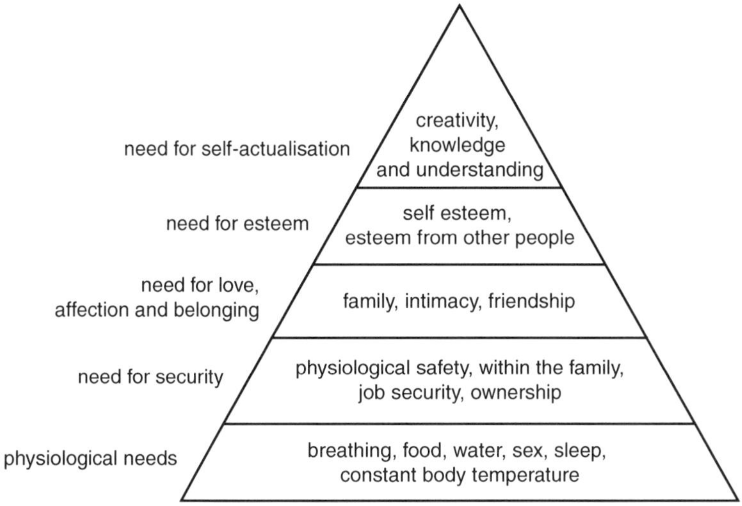

Fig. 1.4 Maslow's Hierarchy of Needs pyramid (adapted Maslow 1954): Physiological needs, need for Security, need for Love, Affection and Belonging, need for Esteem, need for Self-Actualisation (Self-Fulfilment)

Some of them, recurring and generic, can be:

- Feeling of safety;
- Feeling of being useful;
- Feeling of being loved;
- Feeling of being accepted;
- Perception of being in a comfortable environment;
- I can do my work with satisfaction;
- I feel helped;
- I feel part of a group.

Here reference can be made to Maslow (a humanist psychologist) and his Pyramid (Maslow 1954), which gives a hierarchy of human needs (Fig. 1.4). Humanists focus upon potential: they believe that humans strive to reach the highest level of their abilities. Humans seek to break the boundaries of creativity and reach the uppermost levels of consciousness and wisdom. All our basic needs are equivalent to animal instincts (Simons et al. 1987).

Maslow built a hierarchy of five sets of goals which he calls basic needs:

- physiological needs, such as breathing, food, water, sex, sleep, constant body temperature;
- the need for Security, such as physiological safety, security within the family unit, job security, ownership;

- the need for Love, Affection and Belonging, such as the sense of belonging to a family or to a loved on, intimacy, friendship;
- the need for Esteem, this becomes more dominant when the three previous needs are satisfied. This particular need involves both self esteem and the esteem we receive from other people;
- the need for Self-Actualisation (Self-Fulfilment), in other words the desire for knowledge and understanding (Maslow 1954), such as creativity or the need to find and do what we were born to do (Simons et al. 1987).

Let's return to Quality Function Deployment and the three design modes. If QFD is used strictly adhering only to behavioural design, potentially there will be criticism of Quality Function Deployment: it does not seem to support creativity, but rather it seems to lead to "soulless" products. It could be said that QFD is based on explicit requests from the customers yet does not consider the "attractive" ones (you will see in the next chapter the so called Kano's model). Furthermore this method could have limitations in the definition of breakthrough innovations.

But there are good answers to these doubts.

In Chap. 2 we will see the following flow of phases. After market research, using open answer interviews, is conducted in the place where the product and/or service are used (in Japanese *gemba*), the QFD inter-functional team starts a brainstorming process, that consists of meetings, where raw data (phrases, expressions, photos, recordings), collected from the interviews, are organized and reviewed, with the aim of creating new phrases, written following the main rule of "abstraction". We will learn how to use and deploy memos, sticky notes, the same ones used almost everywhere in our offices and homes. The aim of the team is to describe the customer's feelings and emotions.

In this way a working platform is built, ready to be used by engineers, which will give them a better understanding of what the customer wants and how they can meet his requests, without wasting too much energy (and money), on moving in the wrong direction.

This brainstorming activity is completely free and creative.

It is a key phase and will be discussed in detail.

The results, gathered from numerous working groups of students in the university lectures on QFD, confirm that, if correctly applied, *Quality Function Deployment is a creative tool*. I found that, although these groups received the same raw data from customer interviews, the re-worded customer requests differed from each other in some points. This aspect led teams to develop products which were similar in some aspects, yet totally different in others.

Therefore it can be said that from a detailed study of customer requests with QFD we can obtain objects and services that are not identical, but different and excellent, in which Norman's three types of design, visceral, behavioural or reflective, are blended in a definitely creative way.

1.9 The Proposed Framework

The framework that will be discussed from the next chapter on has two aims: educational clarity and practical approach.

This QFD methodology path is mostly sequential and exploits an example of classic bicycle development: the Royal bike.

Also, in Chap. 4, still using the Royal case study, we will see how to manage a kind of mathematics which is no longer classic, i.e. crisp, but is "unfocused", vague, i.e. fuzzy.

All the matrices we will see can be implemented quite easily: in fact common computerised spreadsheets can be used. No specific or expensive software is needed.

The sequence of the activities can be summarised into 7 macro-areas and 12 matrices, see Fig. 1.5. I must underline that not all the framework phases and steps need to be completed in order to obtain some practical results. For instance you can stop the project workflow at a specific macro-area (phase) and, at the same time, you can decide the level of detail and complexity of the calculations, i.e. the dimensions of the matrices involved.

In addiction, "roof" of matrices will not be used. The reader can easily add this analysis to the framework.

Let me list the framework phases:

(a) "Strategy", in other words the evaluation of:

- strategy pursued (matrix 1);
- key core skills (competencies) (matrix 2);
- customer's segments, target for the new product or service to be developed (matrix 3).

(b) "Customer": this is a process of customer analysis that considers two different surveys (the so called *gemba* interviews and a questionnaire survey with closed answers to the questions submitted) and the filling in of an arithmetical matrix, called "Preplan" (matrix 4). The output from this matrix is a weight of the customer's requests, also called Demanded Qualities. This ranking is called Customer Requests priority or Demanded Quality Weight (DQW).

(c) "Product characteristics". In point b) we discussed what the customer or interlocutor want and rank their requests, then QFD activity moves to the area of the product and/or service characteristics, the process analyses them and ranks their priority. Furthermore the method assesses the difficulty in developing performance strong enough to achieve leader status in comparison to competitors; product characteristic bottlenecks are identified, both from the technical and organizational point of view (matrix 5, House of Quality).

(d) "Functions and Mechanisms". Once we have identified and weighted the product characteristics, we move to the functions carried out by the product or

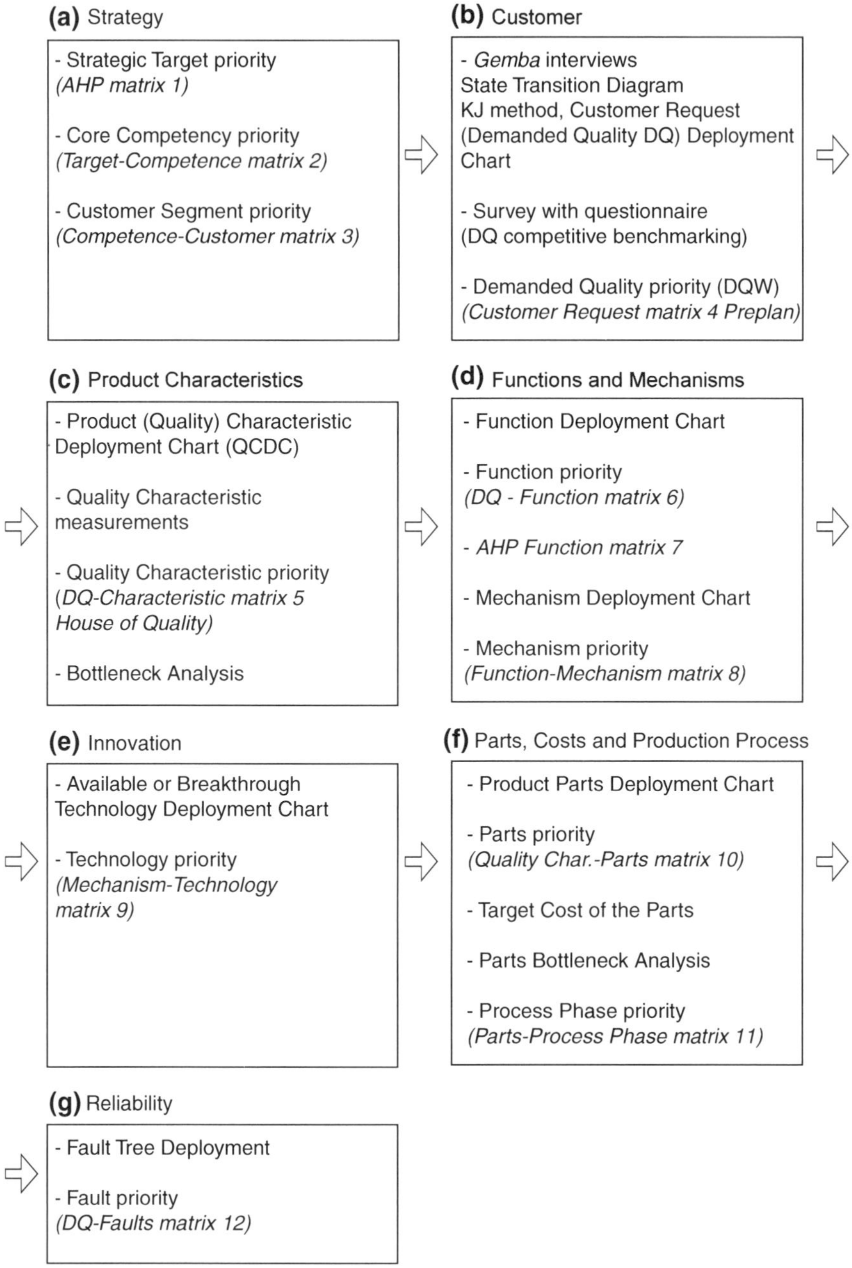

Fig. 1.5 The proposed framework with 7 phases and 12 matrices: Strategy, Customer, Product Characteristics, Functions and Mechanisms, Innovation, Parts, Costs and Production process, Reliability

service, then to the so called "mechanisms", functional components of the product or service. Here again the matrix output gives a ranking weight, a Function priority (matrices 6 and 7) and a Mechanism priority (matrix 8). These weights are very interesting for the engineers because they allow for justified tradeoffs during the design stage.

(e) "Innovation": product innovation can be driven through the deployment of the available and breakthrough technologies. Technologies include available or underdeveloped ones, whereas breakthrough technologies are radical, completely new ones. Technologies are linked to mechanisms to obtain their priority (matrix 9).

(f) "Parts, costs and production process". Some parts are planned as a result of the innovation (point e). Product parts are linked to product characteristics in matrix 10, to get a ranking as a result. With this weight, target costs of the parts can be identified and compared with those calculated by production planning (in the case of internal production) or by suppliers (in the case of external production). In this area we can conduct a bottleneck analysis to identify which parts are cost bottlenecks. These bottlenecks, for example, can be used in the relationships with external parts suppliers. The production process operations are finally linked to parts (matrix 11).

(g) "Reliability". In the end a Fault Tree is developed and linked to the customer's request (matrix 12), because of the effects on customer satisfaction. Fault weights can be used, for example, to manage the FMEA (Failure Mode Effect Analysis) priorities.

References

AIAG (2012) Advanced product quality planning and control plan (APQP). http://www.aiag.org. Accessed 24 Sept 2012

Akao Y (ed) (1990) Quality function deployment (QFD) integrating customer requirements into product design. Productivity Press, Portland

Akao Y, Kogure M, Furukawa Y (1983) Seminar on company-wide quality control and quality deployment. The Arlington Park Hilton, Chicago 31 Oct–3 Nov 1983

Akao Y, Mazur G (2003) The leading edge in QFD: past, present and future. Int J Qual Reliab Manag 20(1):20–35

American Supplier Institute (1987) Quality function deployment. Instruction Manual, ASI, Dearborn

Bellenda R (1992) Il QFD per individuare la qualità richiesta. Sistemi and Impresa (2), pp 58–62, Marzo 1992

Booms BH, Bitner MJ (1981) Marketing strategies and organizational structure for service firms. In: Donnely J, George W (eds) Marketing of Services. American Marketing, Chicago

Booz A (1982) New products management for the 1980s. Hamilton, New York

Camarinha-Matos LM, Macedo P, Oliveira AI, Ferrada F, Afsarmanesh H (2013a) Collaborative environment for service-enhanced products. Institute of Electrical and Electronics Engineers, Jul 29, 2013

Camarinha-Matos LM, Macedo P, Oliveira AI, Ferrada F, Afsarmanesh H (2013b) Supporting product-servicing networks. Institute of Electrical and Electronics Engineers, Oct 28, 2013
Cao Y, Zhao L (2009) FFE Practices of NPD in innovating Chinese manufacturing companies. Institute of Electrical and Electronics Engineers, Jan 7, 2009
Cao Y, Chen R, Zhao L, Nagahira A (2008) Impact analysis of FFE practices of new product development in Japanese companies. Institute of Electrical and Electronics Engineers, Jan 10, 2008
Carnevalli JA, Miguel PC (2008) Review, analysis and classification of the literature on QFD—Types of research, difficulties and benefits. Int J Prod Econ 114(2008):737–754
Carter CF, Williams BR (1957) Industry and technical progress. Oxford University Press, London
Chan LK, Wu ML (2002) Quality function deployment: a literature review. Eur J Oper Res 143 (2002):463–497
Chien TK, Su CT (2003) Using the QFD concept to resolve customer satisfaction strategy decisions. Int J Qual Reliab Manag 20(3):345–359
Clark KB, Fujimoto T (1991) Product development performance. Harvard Business Press, Cambridge
Cohen L (1995) Quality function deployment. How to make QFD work for you. American Supplier Institute, Dearborn
Cooper RG, Project NewProd (1979) Identifying industrial new product success. Ind. Mark. Manage. 8:136–144
Cooper RG (1990) Stage-gate systems: a new tool for managing new products. Bus Horiz 45–54:1990
Cooper RG, Kleinschmidt EJ (1993) Screening new products for potential winners, long range planning
Cristiano JJ, Liker JK, and White CC III (2001) Key factors in the successful application of quality function deployment (QFD). IEEE Trans Eng Manag 48(1):81–95
Fain N, Kline M (2013) The dynamics of multicultural NPD teams in virtual environments. Int J Technol Des Educ 23(2):273–288
Franceschini F (2002) Advanced quality function deployment. Lucie Press, Boca Raton
Fransen Luc (2013) The embeddedness of responsible business practice: exploring the interaction between national-institutional environments and corporate social responsibility. J Bus Ethics 115(2):213–227
Gerstenfeld A (1976) A study of successful projects, unsuccessful project and projects in process in west Germany. ieee transactions on engineering management EM 23(3):116–123
Ginn D, Zairi M, Al-Mashari M, Al-Mudimigh A (2010) Key enablers for the effective implementation of QFD: a critical analysis. european centre for best practice management, research paper: RP—ECBPM/0031, 2010
Govers CPM (1996) What and how about quality function deployment (QFD). Int J Prod Econ 46–47(996):575–585
Govers CPM (2001) QFD not just a tool but a way of quality management. Int J Prod Econ 69 (2001):151–159
Griffin A (1992) Evaluating QFD's use in us firms as a process for developing products. J Prod Innov Manage, 9(3):188–199
Hauser JR, Clausing D (1988) The house of quality. Harvard Bus Rev 66(3):1988
Haverila M (2011) 'Newness to the firm'–variables in the NPD process of technology companies. Int J Bus Innov Res 5(1):29–45
Herstatt C, Verworn B, Nagahira A (2004) Reducing project related uncertainty in the "fuzzy front end" of innovation: a comparison of German and Japanese product innovation projects. Int J Prod Dev 1(1):43–65
King B (1989) Better designs in half the time. implementing quality function in America, 3rd edn. Goal/QCP, Methuen
Lehtinen U, Lehtinen JR (1982) Service quality: a study of quality dimensions, working paper unpublished, service management institute, Helsinki, Finland

Lowe A, Ridgway K (2000) UK user's guide to quality function deployment. Eng. Manag. J. 10 (3):147–155
Lynch P, O'Toole T, Biemans W (2013) From conflict to crisis in collaborative NPD. J Bus Res 67 (6):1145–1153
Mahajan V, Wind J (1992) New product models: practice, shortcomings and desired improvements. J Prod Innov Manag 9:128–139
Martins A, Aspinwall EM (2001) Quality function deployment: an empirical study in the UK. Total Qual Manag 12(5):575–588
Maslow AH (1954) Motivation and personality. Harper, New York
Mazur GH (2002) QFD and design for six sigma: a quality product development system. ISQFD 2002, Munich
McKay J, Marshall P, Grainger N (2014) Rethinking communication in it project management. Institute of Electrical and Electronics Engineers—Jan 6, 2014
Millson MR, Wilemon D (2010) Technology newness as a mediator of NPD strategy, organizational integration, and NPD performance. Institute of Electrical and Electronics Engineers, Jan 18, 2010
Mizuno S, Akao Y (eds) (1994) QFD The Customer-Driven Approach to Quality Planning and Deployment. Asian Productivity Organisation, Tokyo
Mu J, Peng G, MacLachlan DL (2009) Effect of risk management strategy on NPD performance. Technovation, March 2009
Muffatto M, Panizzolo R (1992) Sviluppo di nuovi prodotti. Un modello di riferimento. Sistemi and Impresa 8:17–34
Nishimura K (1972) Ship design and quality table: quality control. JUSE, Tokyo
Norman DA (2004) Emotional design. Basic Books, New York
Pascale RT, Athos AG (1981) The art of Japanese management—applications for American executives. Simon and Schuster
ReVelle J, Moran JW, Cox CA (1998) The QFD Handbook. Wiley and Sons
Rosell DT, Lakemond N (2012) Collaborative innovation with suppliers: a conceptual model for characterising supplier contributions to NPD. Int J Technol Intell Plann 8(2):1–8
Rothwell R (1977) The characteristics of successful innovations and technically progressive firms. R&D Manag 7(3):191–206
Rothwell R, Freeman C, Horsley A, Jervis VTP, Robertson AB (1974) Townsend J sappho updated —project sappho phase. Res. Policy 3(3):258–291
Rubenstein AH, Chakrabarti AK, O'Keefe RD, Soulder WE (1976) Factors influencing innovation success at the project level. Research Management, pp 15–20
Sauer C, Blaize HR (2009) Rethinking IT project management: evidence of a new mindset and its implications. Int J Proj Manag 27(2):182–193
Schonberger RJ (1982) Japanese manufacturing techniques: nine hidden lessons in simplicity, 1st edn. Free Press, New York
Simons JA, Irwin DB, Drinnien BA (1987) Psychology—the search for understanding. West Publishing Company, New York
Sullivan LP (1986) Quality function deployment: A system to assure that customer needs drive the product design and production process. Quality Progress, June 1986
Terninko J (1997) Step-by-step QFD: customer driven product design, 2nd edn. St. Lucie Press Boca Raton Florida
de Poel Van (2007) Methodological problems in QFD and directions for future development. Res Eng Des 18:21–36
Watson GH, DeYong CF (2010) Design for Six Sigma: caveat emptor. Int J Lean Six Sigma 1 (1):19
Xie M, Tan KC, Goh TN (2003) Advanced QFD Applications. American Society for Quality, Quality Press, Welshpool
Yoshizawa T (1997) Origins and development of internationalization of QFD (Japanese). Proceedings of the 6th Symposium on QFD, Tokyo

Zeithaml VA (1981) How consumer evaluation process differ between goods and services. In: Donnelly JH, George WR (eds) Marketing of services. American Marketing Association, Chicago

Zhai L, Hong Z, Zhang R, Nagahira A (2012) The internal mechanism of FFE affecting NPD performance: A theoretical model. Institute of Electrical and Electronics Engineers 8 Nov 2012

Strategic Matrices and Customer Analysis

2

Abstract
In this chapter the first two phases of the QFD framework will be discussed:

- Strategy;
- Customer.

In the QFD "strategy" area, we will calculate:

- the priority of strategic targets;
- the priority of our core competencies;
- the priority of the customer segments.

Priority is a numerical value we obtain as an output of a matrix such as:

- Analytic Hierarchy Process (AHP) Matrix;
- QFD correlation matrix;
- Preplan matrix.

Displaying customer segment priorities and their deployment with a graph will allow for focused planning of the first activity around the customer: the open answer "*gemba*" interviews with a small group of selected customers. After collecting several raw data, phrases and expressions from the customers interviewed, the QFD team will start brainstorming sessions making use of the Jiro Kawakita (KJ) method, to display, organise and select ideas and data. The output of KJ is a set of sentences that form the so called Demanded Quality Deployment Chart. A QFD questionnaire will then be prepared and sent to a large group of customers, contacts and potential customers. Following this questionnaire competitor benchmarking starts. The average scores obtained from the questionnaire will be inserted in a matrix called "*Preplan*" the output of which will be an assessment of the customer's requests (Demanded Qualities) through an

D. Maritan, *Practical Manual of Quality Function Deployment*,
DOI 10.1007/978-3-319-08521-0_2

index called Demanded Quality Weight. An educational case study (the development of "Royal" classic bicycle) is used to describe each step of the method. Exercises for the reader are proposed at each stage.

2.1 Introduction

We have seen that one of the greatest difficulties we meet, when we try to apply QFD, lies in the sensation of confusion we have when we read scientific and didactic publications about the topic. These often focus on specific aspects of the method or they examine its managerial and performance effects. Furthermore one of the strong points but also a terrible trap is the possibility to modify it through the use of different links, files, charts, diagrams, which vary from context to context. The innumerable scientific publications have taken the calculation procedures to an extreme level, and they have highlighted potentialities of QFD on the most disparate areas of interest (Chap. 1 deals with some scientific reviews), but they have presented very few clear, and, especially, complete examples, that are also comprehensible to most people, which could allow someone wanting to start using QFD, to follow a sufficiently simple sequence of activities and achieve effective results.

In this chapter we begin to deploy the framework presented in chapter one, starting from the first thing that must be done: understanding our strategic targets, our strengths and the customers.

2.2 What You Will Learn in This Chapter

In this chapter the first two phases of the framework will be discussed (Fig. 2.1):

(a) Strategy;
(b) Customer.

We will spend more time in these areas, compared to the others, because it is also necessary to illustrate what calculations should be done in the QFD matrices and how the team meetings should be managed during brainstorming sessions.

It is recommended that whilst reading this book frequent reference is made to this section, so that the sequence of the activities that need to be carried out can be memorised and there will never be a feeling of being "lost in fog". Furthermore, if you are thinking of your own QFD project, remember that the term product is intended as a complex mixture of both service and the physical object and that the term customer as a client in a business transaction or, more generally, your interlocutor.

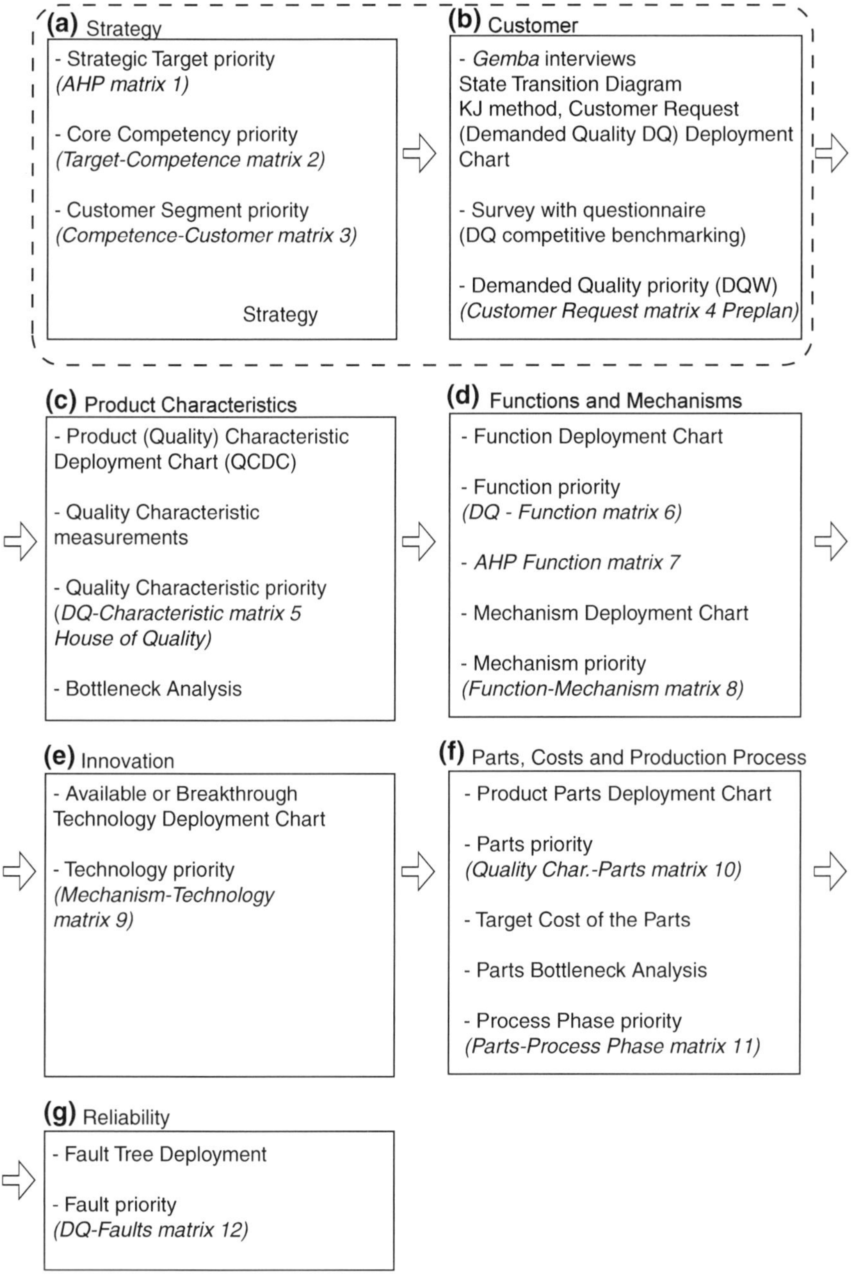
(a) Strategy
- Strategic Target priority
(AHP matrix 1)
- Core Competency priority
(Target-Competence matrix 2)
- Customer Segment priority
(Competence-Customer matrix 3)
Strategy
(b) Customer
- Gemba interviews
State Transition Diagram
KJ method, Customer Request
(Demanded Quality DQ) Deployment
Chart
- Survey with questionnaire
(DQ competitive benchmarking)
- Demanded Quality priority (DQW)
(Customer Request matrix 4 Preplan)
(c) Product Characteristics
- Product (Quality) Characteristic
Deployment Chart (QCDC)
- Quality Characteristic
measurements
- Quality Characteristic priority
(DQ-Characteristic matrix 5
House of Quality)
- Bottleneck Analysis
(d) Functions and Mechanisms
- Function Deployment Chart
- Function priority
(DQ - Function matrix 6)
- AHP Function matrix 7
- Mechanism Deployment Chart
- Mechanism priority
(Function-Mechanism matrix 8)
(e) Innovation
- Available or Breakthrough
Technology Deployment Chart
- Technology priority
(Mechanism-Technology
matrix 9)
(f) Parts, Costs and Production Process
- Product Parts Deployment Chart
- Parts priority
(Quality Char.-Parts matrix 10)
- Target Cost of the Parts
- Parts Bottleneck Analysis
- Process Phase priority
(Parts-Process Phase matrix 11)
(g) Reliability
- Fault Tree Deployment
- Fault priority
(DQ-Faults matrix 12)

Fig. 2.1 Phases **a** and **b** of the framework, Strategy and Customer

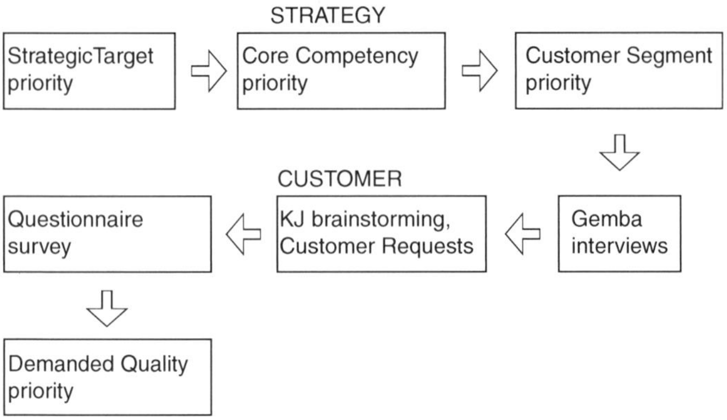

Fig. 2.2 Workflow of this chapter, Strategic Target priority, Core Competencies priority, Customer Segments priority, *gemba*, KJ, survey, Demanded Qualities priority

Below, some questions to be answered.
For phase (a):

- What new product development strategy are we looking for?
- What are our core competencies, our strengths?
- Who are our customers, who is the new product designed for?
- Who will carry out these phases?

For phase (b):

- Where can we get information about our customer from?
- How do we manage the interview with the customer? How is a questionnaire created?
- How do we arrange the raw data that the customer gives us?
- How do we rank the priority of a customer's request? How should this priority be displayed? How do we read it in practical terms?

See Fig. 2.2 for the workflow of this chapter. The aim is to answer the previous questions about the strategic approach we want to follow and the customer. Three kinds of matrices will be used:

- Analytic Hierarchy Process (AHP) Matrix;
- QFD correlation matrix;
- Preplan matrix.

In the "strategy" area, we will calculate:

- the priority of strategic targets;
- the priority of our core competencies;
- the priority of the customer segments.

Priority is a numerical value we obtain as an output of a matrix. We can also call this priority "weight" or "importance".

Displaying customer segment priorities and their deployment with a graph will allow for focused planning of the first activity around the customer: the open answer "*gemba*" interviews with a small group of selected customers. After collecting several raw data, phrases and expressions from the customers interviewed, the QFD team will start brainstorming sessions making use of the Jiro Kawakita (KJ) method, to display, organise and select ideas and data.

The output of the brainstorming using the KJ method is a set of written sentences that form the so called "Demanded Quality Deployment Chart". A questionnaire will then be prepared and sent to a large group of customers, contacts and potential customers. Following this questionnaire, competitor benchmarking starts. The average scores obtained from the questionnaire will be inserted in a matrix called "Preplan" the output of which will be an assessment of the customer's requests (Demanded Qualities) through an index called Demanded Quality Weight.

Let's start with the case study that will be used to illustrate the framework practically.

2.3 The Royal Classic Bicycle Case Study

During lectures, when the concept of strategy in Quality Function Deployment is presented to students, who are maybe listening distractedly because they expect the lecture to be boring, some provocative questions can be asked.

What aims are you pursuing with your studies? Could you list some of them? Which of them are more important than the others? Secondly, have you evaluated your personal, specified strengths? Could you list some of these? Do you have perseverance or creativity? Dialectic skills? What kind of profession do you think you'll go into? Which factories or companies are you thinking of sending your applications to? In the end, who are your clients?

The reader can imagine that all these questions will always wake even the sleepiest in the lecture hall. The students change position from lying across the desk to sitting with their arms folded on the table. Taking the university student careers as an example, we always find that they are unable to give immediate answers to these simple banal questions:

- targets for the end of a difficult course, which often lasts much longer than expected;
- subjective skills;
- type of company to send the curriculum to.

Everyone thinks that these concepts are clear, well-defined, assimilated, but instead we find that they are very tiring to write and list, and the results are not guaranteed at all.

This example of provocation causes both confusion and interest yet certainly leads listeners to greater reflection on the matter.

Whether it is for a university course, or to start a new company or for the development of a new product, Quality Function Deployment makes us think, play, sketch, write, draw and express complex concepts in a structured way. QFD is not only useful for its numerical results but also and particularly for its apparently slow process of systematic analysis. In this sense it can be considered, as mentioned previously, one of the best tools to support decisions in complex environments.

So let's move to phase (a) of our framework.

An educational case study (the development of a classic bicycle, called "Royal") is used to describe the method.

The classical/vintage bicycle market (see Fig. 2.3 for the main components of a bicycle) is typically a niche market and, in Europe, it consists of bicycles with traditional frames, painted in sober colours (black, brown, ivory), equipped with luxurious gadgets, often made of leather. On the other hand, in the USA classical bikes have generously curved frames, with large wheels and a lot of chrome. Here, however the type we refer to is a European style product.

The organizational chart of the company, that is going to develop "Royal", is functional and it is deployed in various divisions: marketing, logistics, production, human resources, and quality offices. There is also a development office, that follows the development of a product starting from market analysis and going through product planning, styling, engineering, prototypes and testing to the preproduction phase.

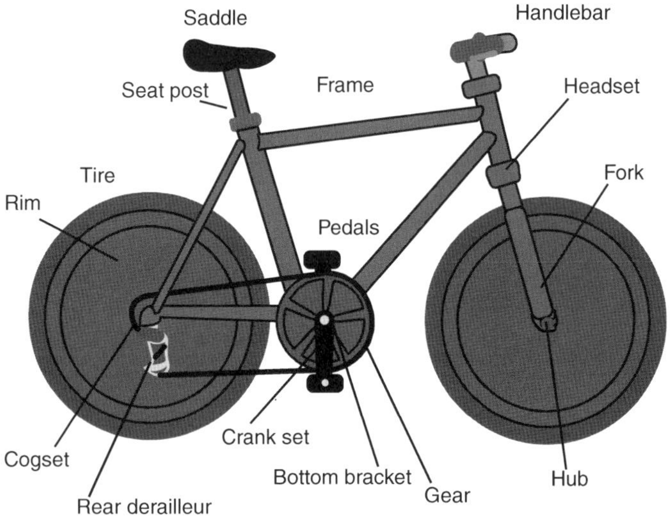

Fig. 2.3 Main components of a bicycle

The QFD project, presented in these chapters, refers to the development of a model of bicycle featuring classic style, historic fashion and marketed under a brand with a reputation for excellent products, handcrafted components, perfection in design, to the extent of extreme focus on high quality details.

A historical brand, the challenge of a company: everything stems from the need to innovate a line of products, which are objects of an almost obsessive appreciation by some customers. Input from top management is clear: the style must remain classical and conservative, prestige must be reinforced and also, due to imperfect management of the product over the last few years, there have been criticisms about the quality of the components and the finishing of the product, seen by distributors, bike shops and customers as not being in line with the fame of the brand; in contrast, customers also seem to demand a moderate price.

Top management guidelines, decided during a meeting extended to the company owners, are to upgrade (technical restyling) a model that is traditionally considered by clients as the perfect expression of the brand: the "Royal" bicycle.

The mission is clear: to seek the highest quality and excellence with no compromises, yet at the same time developing a "saleable" product.

2.4 Strategy Deployment

As touched on in the introduction, QFD requires, first of all, suitable knowledge and awareness of the strategic objectives, the company's internal competencies and the customer segments that the project is dealing with.

From a practical point of view, in brainstorming meetings the members of the QFD workgroup meet together to discuss different opinions about the strategic subject matters, guided by the need to fill out three matrices, that can be called "Strategic matrices" (Fig. 2.4).

The first matrix compares the strategic targets in pairs, using a method called "Analytic Hierarchy Process" (AHP); unlike the first, the second and third matrices compare sentences and verbal expressions, belonging to different contexts; in this case the symbolism and the technique of calculation, as we will see, are developed with a classical QFD style.

It is important to understand that the speed, at which rules of market competitiveness change, is so high that we need to consider a QFD analysis over short spans of time.

Quality Function Deployment assists medium or short term product planning.

Therefore these matrices must be periodically re-examined, because the targets, the core competencies, and the customer segments tend to change very rapidly.

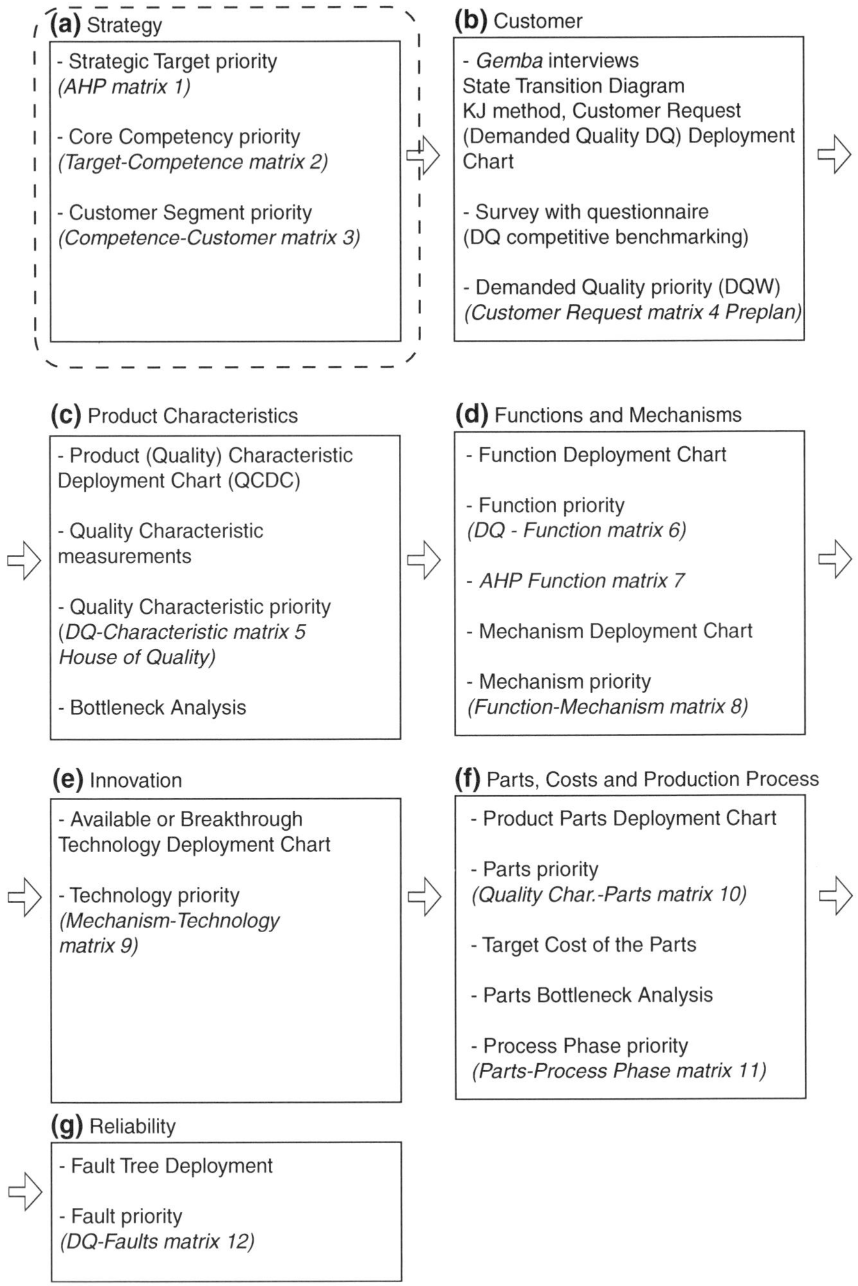
(a) Strategy
- Strategic Target priority
(AHP matrix 1)
- Core Competency priority
(Target-Competence matrix 2)
- Customer Segment priority
(Competence-Customer matrix 3)
(b) Customer
- Gemba interviews
State Transition Diagram
KJ method, Customer Request
(Demanded Quality DQ) Deployment
Chart
- Survey with questionnaire
(DQ competitive benchmarking)
- Demanded Quality priority (DQW)
(Customer Request matrix 4 Preplan)
(c) Product Characteristics
- Product (Quality) Characteristic
Deployment Chart (QCDC)
- Quality Characteristic
measurements
- Quality Characteristic priority
(DQ-Characteristic matrix 5
House of Quality)
- Bottleneck Analysis
(d) Functions and Mechanisms
- Function Deployment Chart
- Function priority
(DQ - Function matrix 6)
- AHP Function matrix 7
- Mechanism Deployment Chart
- Mechanism priority
(Function-Mechanism matrix 8)
(e) Innovation
- Available or Breakthrough
Technology Deployment Chart
- Technology priority
(Mechanism-Technology
matrix 9)
(f) Parts, Costs and Production Process
- Product Parts Deployment Chart
- Parts priority
(Quality Char.-Parts matrix 10)
- Target Cost of the Parts
- Parts Bottleneck Analysis
- Process Phase priority
(Parts-Process Phase matrix 11)
(g) Reliability
- Fault Tree Deployment
- Fault priority
(DQ-Faults matrix 12)

Fig. 2.4 Framework phase **a**, Strategy deployment

2.4.1 Matrix 1. Strategic Targets Priority and AHP Method

We know that the optimal composition of the team working with Quality Function Deployment is made up of experts that come from different offices, i.e. sales, marketing, engineering, quality, logistics and production. However, in this initial phase those at the highest levels of the company have to be involved, the owners, for example, the CEO, the managing director, the sales managers.

Their task is to work thoughtfully and to express the "Strategic Targets" of the QFD project their company is starting on.

These targets should be made up of about four up to a maximum of ten sentences, one for each different target, and constructed in such a way as to be clear and understood by all the members of the group.

Arranging of similar targets in groups, if needed, must be managed on the basis of subjective affinities, arranged by the QFD team.

The concept of an agreed selection of targets is important, so that the process of constructing these sentences is an opportunity for top management to stop and meditate about the product development process strategies.

In the "Royal" case study the strategic targets identified are:

- Reinforcement of the market share;
- Re-launch of the brand;
- Increase in turnover;
- Reduce the time to market (the time for product development up to its market launch).

Different levels of priority are now assigned to these objectives using Analytic Hierarchy Process (AHP). Analytic Hierarchy Process (Saaty 1980; Mazur 1996) is a technique for supporting decisions, developed by Saaty during the Seventies, in order to work with complex problems of a technological, economical and socio-political nature (Ramadhan et al. 1999). The objective of AHP analysis is to determine the best option (or a rank), among those available to the decision maker by studying the subjective importance that they each have compared to the others.

Having identified the strategic targets of the "Royal" project, we can organize them traditionally in a list of decreasing importance and simply consider them all at the same time. This approach proves easy only when we have few objectives. Or we can (and we normally do) write them down on a piece of paper and then put these in order starting from the most important target, the second most important one, the third and so on, up to the sentence with the lowest priority.

Unfortunately, as the number of sentences increases, this process becomes too approximate. Instead, it is easier and more reliable to compare the targets in couples, focusing on a simple, specific question, without considering the entire list. For example just be asking ourselves is "sentence x" more or less important than "sentence y" and how much more important is it?

AHP helps us do this.

AHP STRATEGIC TARGET MATRIX 1

1 = the row is equally important as the column
3 = slightly more important
5 = more important
7 = much more important
9 = very much more important
0,33 = slightly less important
0,20 = less important
0,14 = much less important
0,11 = very much less important

STRATEGIC TARGET DEPLOYMENT		STRATEGIC TARGETS: A Reinforcement of the market share	B Relaunch of the brand	C Increase turnover	D Reduce the time to market	a (Normalised assessment)	b	c	d	Absolute weight	Strategic target priority
1	Reinforcement of the market share	1.00	0.20	5.00	3.00	*0.15*	*0.12*	*0.31*	*0.41*	0.99	**24.87%**
2	Relaunch of the brand	5.00	1.00	7.00	3.00	*0.77*	*0.60*	*0.44*	*0.41*	2.21	**55.28%**
3	Increase turnover	0.20	0.14	1.00	0.33	*0.03*	*0.08*	*0.06*	*0.05*	0.22	**5.55%**
4	Reduce the time to market	0.33	0.33	3.00	1.00	*0.05*	*0.20*	*0.19*	*0.14*	0.57	**14.30%**
	Total	6.53	1.67	16.00	7.33	*1.00*	*1.00*	*1.00*	*1.00*	4.00	100.00%

Fig. 2.5 Royal AHP Strategic Targets matrix 1

You can use any software for spreadsheets. Then, build a matrix in which the first row and the first column are neatly filled by the targets sentences. Figure 2.5 illustrates the matrix 1 for the Royal project. Let's now see the calculation steps in sequence.

(1) In every cell of the matrix (area on the left side) a numeric value is inserted, chosen in a predetermined range that corresponds to the assessment of how important the specific objective row is compared to the corresponding objective column. Remember the flow direction: row versus column, not the opposite. The range of values used is the following:

1 = the row is equally important as the column
3 = the row is slightly more important than the column
5 = more important
7 = much more important
9 = very much more important
0.33 = slightly less important
0.20 = less important
0.14 = much less important
0.11 = very much less important.

For example the target "Reinforcement of the market share" is considered by the team "less important" than the objective "Re-launch of the brand" and so the value 0.20 is inserted in the relationship cell.
We see that in changing the term from "less" to "more", the weight is inverse to the other; for example, "quite more important" corresponds to a weight of 5 that

is equal to 1/0.2. All values in the diagonal of the matrix are "1.00", because every objective is compared to itself and the sentence "equally important" has to be used. Finally the lower triangle on the left is made up of values that are the inverse of those in the upper triangle on the right of the matrix; indeed the two targets are inverted in the order of comparison.

If we consider the couples of targets, we have to specify only a part of the connections between them: the cells that need to be filled are those situated in the upper part on the right side of the diagonal of the matrix, coloured light grey: in the case of four targets, the number of relationships to be determined becomes six.

(2) The second step is to normalise our assessments along the columns of the matrix: first, add together the values in the specific column of the target (group A–D), second, divide the value of every cell by the sum of the respective column; on the right side of the initial matrix, there are a number of normalised columns (a, b, c, d in Fig. 2.5), equal to the number of the targets. For example in cell "a1" the calculated value is:

$$\frac{A1}{\sum A_j} = 0.15$$

(3) Now, in the following stage, we sum the normalised values, this time along the rows: we obtain the so-called "Absolute weight" of the strategic targets, which is shown in the penultimate column of Fig. 2.5, on the right.

(4) Finally, for every objective, the "relative" value of the absolute weight is the "*Strategic Target priority*", calculated dividing the absolute weight by the sum of the absolute weights (that is the number of targets, because of normalisation), represented as percentage.

The results of the calculation, namely the strategic targets weights, listed in decreasing order and represented in a simple graph (Fig. 2.6). This graph helps us consider the trend of the relative weights of the targets and to identify groups, characterised by similar priorities. I suggest using planar histograms and uniform colours; furthermore, the y-axis might have to be "cut" at its lower point (in this case it doesn't start from a zero origin), in order to "separate" the values of the items and to represent the different priority better visually.

Let's now comment a little on the Royal target priority histogram.

Firstly, we can note that the trend is exponentially decreasing. This characteristic is also common in the next matrices. This trend is useful in decision making situations because it is easier to take a decision if there is a non linear assessment.

We can observe that the target "Re-launch of the brand" has the greatest priority and that it is of much greater importance than the two intermediate targets, "Reinforcement of the market share" and "Reduce the time to market"). "Increase

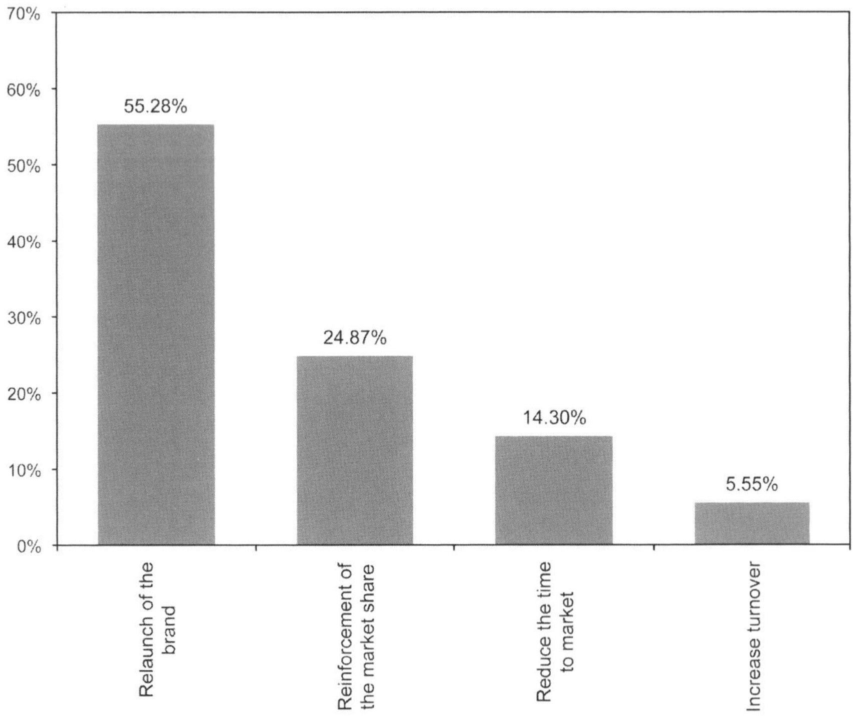

Fig. 2.6 Histogram of the Royal Strategic Target priority

the turnover" has the lowest priority. One has to admit that this smart way of representing the data is extremely useful and should not taken for granted. From my own personal experience, none of the companies I have worked with as consultant or teacher was able to show, in answer to a precise request, what its product developing targets were and/or their priority in this detail.

We humans often forget to do the simple things.

Exercise Focus on a simple developing project, which concerns you or your company. Ask yourself or top management (if possible) if you can make a written list of strategic targets and to establish their priority.

Find and deploy 4 targets you have for your own project and create the AHP matrix. You can use a spreadsheet in LibreOffice, OpenOffice or Microsoft Office(©) suites.

Sort the relative weights in descending order and construct a histogram like Fig. 2.6. Take time to comment on the results. Do some of the target weights surprise you?

2.4.2 Matrix 2. Core Competency Priority and the Independent Scoring Method

The phase after the strategic targets focuses on the inner (also called "core" or "strategic") competencies and on the impact that they have to reach the targets.

Core Competencies are "the collective learning in the organization, especially how to coordinate diverse production competencies and integrate multiple streams of technologies" (Prahalad and Hamel 1990). They are the foundation on which a company (or a person) creates competitive advantage in society. Or in other words core competencies are those things that a company or an organisation can do better, their strengths. For example, we can say that in the automotive industry a strategic core competency in Volvo is safety and a strategic core competency in Toyota is reliability. According to Prahalad and Hamel, a core competency has three characteristics:

- it allows for potential access to a wide variety of markets;
- it increases benefits perceived by the customer;
- it is difficult for the competitors to imitate.

The core competencies also represent "combinations of input and knowledge-based resources that exist at higher levels in a hierarchy of integration" (Galunic and Rodan 1998). They are the medium by which all the potential of a company or organisation adapt to the variations of the surrounding environment.

In this second phase the team of top managers' assignment is to identify the company's core competencies.

It is suggested a limited number of competencies, variable from 4 to 10.

They are then expressed with descriptive, clear, agreed sentences.

In the case of the Royal bicycle development, the core competencies identified are:

- We know our customers;
- Widespread distribution network;
- Technical experience and specific bicycle know-how;
- Good financial capacity;
- Good market share for classic bikes.

Now it is time to begin the study of the capacity that the various internal competencies have to achieve the strategic targets, with AHP.

A new matrix calculates their order of importance, their priority. As we'll discover, it looks like the typical one in Quality Function Deployment, with the use of symbols and mathematics that you will get to know very well as you read this book (Fig. 2.7).

The first column on the left of the matrix contains the list of the previously identified Strategic Targets. They are the rows of this matrix.

STRATEGIC TARGET - CORE COMPETENCY MATRIX 2								
⊙ : STRONG CORRELATION ⦸ : MEDIUM CORRELATION ⁂ : WEAK CORRELATION			CORE COMPETENCY DEPLOYMENT					
			A	B	C	D	E	
STRATEGIC TARGET DEPLOYMENT		Target priority	We know our customers	Widespread distribution network	Technical experience and specific bicycle know-how	Good financial capacity	Good market share for classic bikes	
1	Reinforcement of the market share	24.87%	⦸	⁂		⦸	⁂	
2	Relaunch of the brand	55.28%	⊙	⦸		⊙	⦸	
3	Increase in turnover	5.55%	⁂	⁂		⁂		
4	Reduce the time to market	14.30%		⁂	⊙	⁂		Total
		Absolute weight	5.78	2.11	1.29	5.92	1.91	17.00
		Core competency priority	**33.99%**	**12.39%**	**7.57%**	**34.83%**	**11.22%**	100.00

Fig. 2.7 Royal Strategic Target-Core Competency matrix 2

Next to the targets, the second column on the left contains the Strategic Targets weights, the output of the AHP matrix.

Core Competencies are placed as headers of the following columns.

In this way, the resulting matrix is made up of:

a number of cells equal to the product between the number of targets and the number of competencies.

The working group has to examine, cell by cell, the current connection between core competencies and targets. For each cell, the team has to detail if some correlation exists between each core competency and each strategic target. Symbols of correlation can be inserted or the cell can be left empty should there be a lack of correlation. If we remember what we read in Chap. 1, we recognise this structure is similar to House of Quality (Sect. 1.6).

In this case Whats are the Strategic Targets and Hows are the Core Competencies.

For this matrix, note that the flow direction is "column" to "row".

The symbols generally used are:

⊙ = strong correlation
⦸ or ○ = medium correlation
⁂ or Δ = weak correlation.

I suggest using software spreadsheets to build all the QFD matrices needed. For this second matrix, the font Wingdings 2 (© Microsoft), or similar, and the equivalent letters "8", "W", "ù" are very useful.

According to Akao (1990), the weights corresponding to symbols are respectively:

"9" for strong correlation,
"3" for average correlation and
"1" for weak correlation,
the empty cell corresponds to weight "0".

We use w_j to indicate the absolute importance (weight) of every competency. For every competency (for every column) absolute weight is calculated by adding the products of the strategic target priority (the second column) and the weight of the corresponding symbol in the same row.

We have vector:

$$w_j = \mathbf{d} * R_j$$

where:

d Strategic targets weights vector-row;
R_j column j of the matrix.

This type of calculation is called "Independent Scoring Method" (Mitsufuji in Akao 1990), which can be abbreviated as "ISS".

As an example, the absolute weight of strategic competency "We know our customers" is calculated in the following way:

"We know our customers" Absolute weight =

$24.87\,\% * 3 + 55.28\,\% * 9 + 5.55\,\% * 1 + 14.30\,\% * 0 = 5.78.$

The absolute weight finally evolves to the relative value, the so-called "Core Competency weight" or Core Competency priority, obviously calculated by dividing the single absolute weight by the sum of absolute weights; in this way the weight of the competency "We know our customers" is achieved:

"We know our customers" priority = (5.78/17.00) = 33.99 %.

Again, as in the targets AHP matrix, the weights of the competencies can be sorted in descend order and inserted in a histogram (Fig. 2.8).

Let's now comment on company core competencies priority distribution for the Royal project. You can see that the trend is still exponentially decreasing. Two competencies have high relative priority: "Good financial capacity" and "We know our customers". The "Technical experience and specific bicycle know how" competency seems to have low priority (importance) for achieving the targets predetermined by the project.

This second phase is very interesting for decision makers, the top management of a company, the owner-manager of a small office, the staff of an organisation or a self-employed person.

The time and the economical resources dedicated to these calculations are generally moderate. Two or three hours for a four-five core competencies matrix could be enough if the spreadsheet is already software developed.

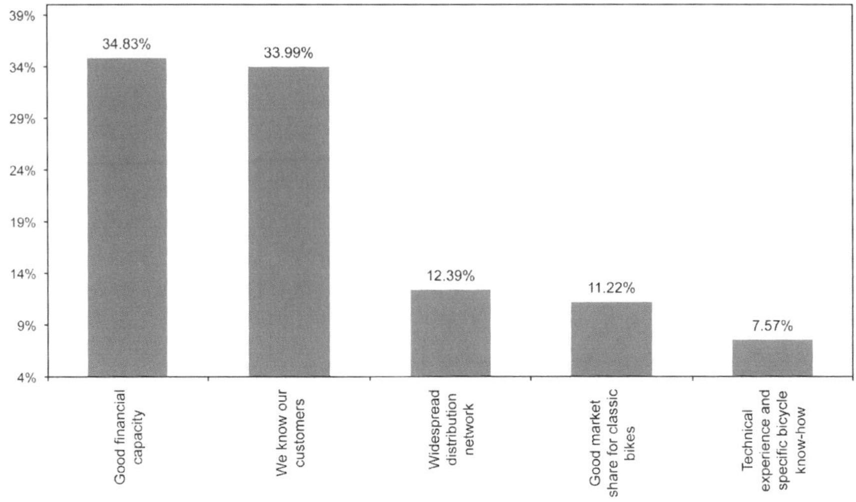

Fig. 2.8 Histogram of the Royal Core Competency priorities

Exercise Firstly ask yourself or top management (if possible) if you can make a written list of core competencies and establish their priority. Are these core competencies communicated to employees in your company?

Find and deploy 4 core competencies you or your company have and create the targets-core competencies matrix 2. You can use a spreadsheet in LibreOffice, OpenOffice or Microsoft Office (©) suites. Sort the relative weights in descending order and construct a histogram like Fig. 2.8. Take time to comment on the results. Ask a colleague or a friend, who knows your project to fill the same matrix with symbols. What happens? Are the results the same as yours? Why?

2.4.3 Matrix 3. Customer Segment Priority

Let's go to the last phase of section (a) of our framework which focuses on Clients.

The team must identify the main Customer segments.

It is possible to use consolidated segmentation of the specific sector or the standard type of customer as defined by marketing research institutes through their periodical market analysis. There are different types of customers a company speaks to. A customer can simply be someone who buys the product, as for retail companies who deal with the end user, or other companies if we consider the industrial market. We can even think about the customer within the same company organisation.

To support fragmentation we can create a chart divided into 6 columns called "*Customer Segment Table*". They represent the 5W1Hs questions, i.e. Who (who will use the product or the service), What (what they will be used for), When (when, on what occasion), Where (where they will be used), Why (for which reason), How (how the service or the product will be used).

As for the previous two matrices, I suggest a limited number of segments so as not to make calculation too difficult. We are talking about matrices, which should be defined by top management and therefore should be simple, lean and essential.

In the case of the Royal classic bicycle top managers identify four types of customer:

- Specialised retailer;
- Large scale retail trade;
- Adult end user;
- Young end user.

The third matrix has the same configuration and layout as the second one. In this case headers and lines are made up of core competencies while columns indicate the segments of customers. In the column, on the right side of the competencies list, we write the results of the second matrix, i.e. their relative weight (core competencies priority).

But what logic should we follow when we fill in the internal cells?

In this case we must evaluate how different internal competencies affect the Customer types: in the third matrix, therefore, we read from "line" towards "column".

We also use ISS (Independent Scoring Method) to determine the absolute weight of the customer segments (Fig. 2.9).

The absolute weight is then made relative and the results, as for the competencies evaluation, are represented in a histogram (Fig. 2.10).

"Specialised retailer" is the segment obtaining highest priority, followed by "Adult customer". "Large scale retail trade" does not represent a strategical area for the development project of the "Royal" classic bicycle.

Exercise Firstly ask yourself or top management (if possible) if you can make a written list of customer typologies and to establish their priority. Is this segmentation communicated to employees in your company?

Find and deploy 4–5 customer segments for your project and create the core competencies—customer segments matrix 3. You can use the same spreadsheet you developed for matrix 2. Sort the relative weights in descending order and construct a histogram like Fig. 2.10. Take time to comment on the results. Are there any surprises compared with the usual segmentation priority that your marketing office has? Why? Are there any errors in the matrix completion?

Take time to review the three strategic matrices workflow.

CORE COMPETENCY - CUSTOMER SEGMENT MATRIX 3							
⊙ : strong correlation ⊘ : medium correlation ∴ : weak correlation			Customer				
			1A	2B	3C	4D	
	CORE COMPETENCY DEPLOYMENT	Core competency priority	Specialised retailer	Large scale retail trade	Adult end user	Young end user	Total
A	We know our customers	33.99%	⊙		⊙	∴	
B	Widespread distribution network	12.39%	⊙	∴	⊘	⊘	
C	Technical experience and specific bicycle know-how	7.57%	⊘	∴	⊘	∴	
D	Good financial capacity	34.83%	⊙	⊘	⊘	⊘	
E	Good market share for classic bikes	11.22%	⊙	∴	⊙	∴	
		Absolute weight	7.54	1.24	4.70	1.83	15.32
		Customer segment priority	**49.20%**	**8.13%**	**30.71%**	**11.96%**	100.00%

Fig. 2.9 Royal Core competency—Customer Segment matrix 3

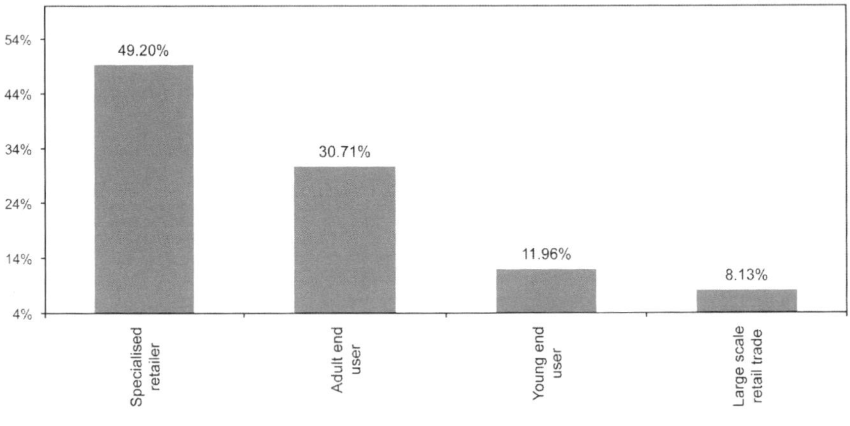

Fig. 2.10 Histogram of the Royal Customer Segment priority

2.5 Customer Analysis. From "Gemba" to the Demanded Qualities (DQ)

After setting the strategical analysis, which clarifies objectives, competencies and customer segments, how they are linked and their priority, our Quality Function Deployment framework concentrates on Customer. A sequence of steps can be followed, divided into:

- on site interviews and survey;
- data management through calculation matrices.

The requests of the customer, also called "Demanded Qualities" (DQ), will be identified thanks to a number of open interviews (the so-called *gemba*) and a new organisation of the qualitative sentences collected (Jiro Kawakita method, KJ). The Demanded Qualities, that customers use to describe their needs and expectations, will be used to plan a questionnaire with closed answers, which will supply the numeric basis for a calculation matrix called *"Preplan"* matrix by Akao (1990). The final target of this matrix is to calculate the priority of the Customer Requests (also called Demanded Quality Weight, DQW, Fig. 2.11).

2.5.1 Gemba Analysis

Analysis of Customer Requests is a fundamental point in the QFD method.

It forces the working team to re-think and re-work the company's idea, often taken for granted, of the customers' behaviour, buying habits and the use they make of the product. Most of the time this idea is obsolete or does not correspond to reality. It is sometimes a hard blow for the sales organisation and for the technical department, but it is a tonic granting the company survival.

We start from two Japanese ideograms that form the word "*gemba*": it means "the place where truth is known", "the actual place". In other words, it is the "source", indicating a location (Mazur 1999, Ronney et al. 2000). In Japan, the crime scene is called *gemba* by detectives, and journalists during TV news reports say they are speaking from *gemba* (Wikipedia 2009). In economics the word *gemba*, "the real source of information", is the place where the product or service acquires its value for the customer, where it is actually used. For industry it is the production area; *gemba* can also express the idea of show rooms, shops or where the service interacts with the user of the goods (Masaaki 1997).

QFD *gemba* analysis is made up of open answer interviews, addressed to customers or potential customers, while they are using the product/service; they are dialogues between supplier and customer, where the latter is led to give his opinion, make suggestions, criticise, and judge the product or service, and make a comparison with the competitors.

Is *gemba* too different from traditional market research methods?

Yes. Using traditional methods, products and services are developed inside the companies and afterwards prototypes are offered to customers who are then asked for their evaluation. There can also be focus groups or telephone interviews, whose purpose is to measure how attractive the new product is or how good its performance is. Mazur (1999) considers *gemba* as the place where customers actually use the product. The analysis therefore is qualitative instead of quantitative, and it is made before the product development, putting user sensations and opinions first.

Watching customers "at work" is an excellent way to learn more (Mazur 1996).

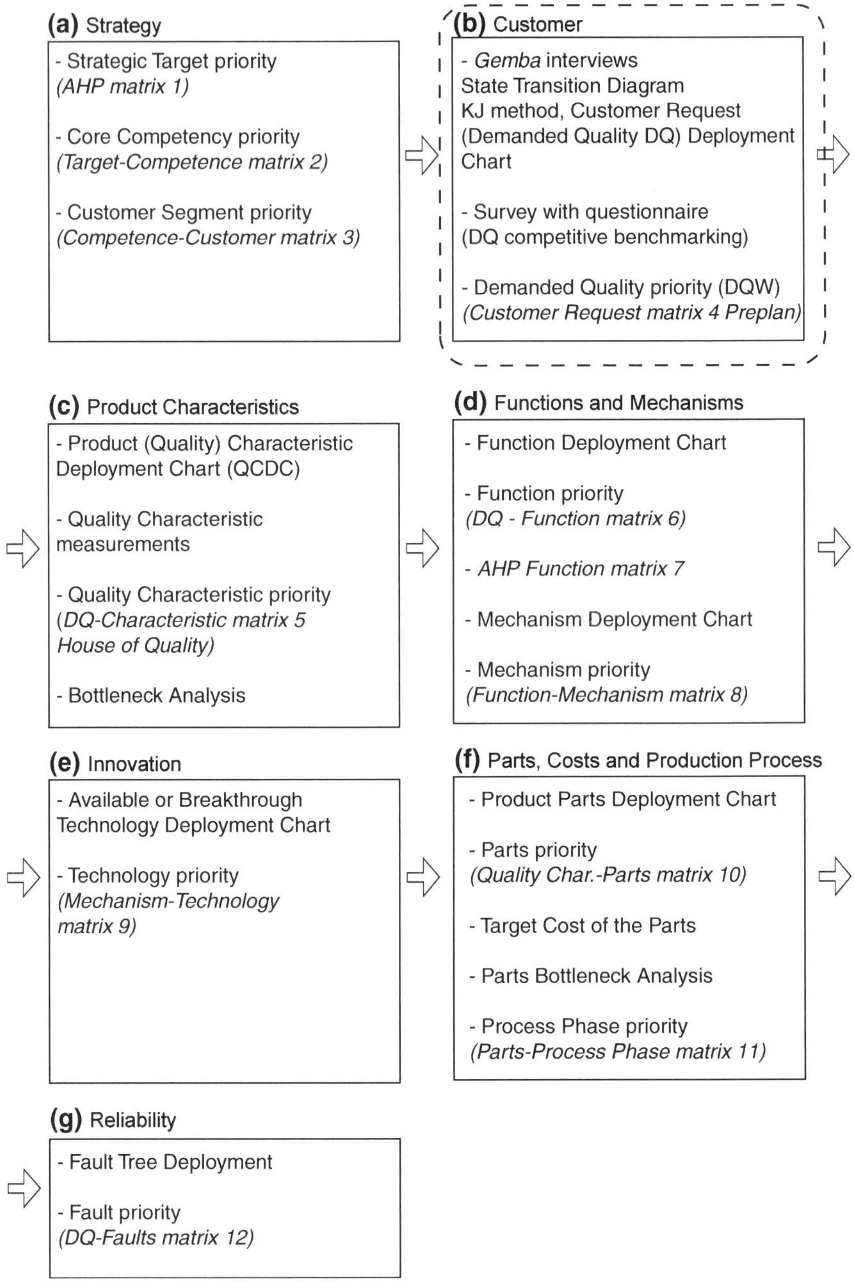
(a) Strategy
- Strategic Target priority
(AHP matrix 1)
- Core Competency priority
(Target-Competence matrix 2)
- Customer Segment priority
(Competence-Customer matrix 3)
(b) Customer
- Gemba interviews
State Transition Diagram
KJ method, Customer Request
(Demanded Quality DQ) Deployment
Chart
- Survey with questionnaire
(DQ competitive benchmarking)
- Demanded Quality priority (DQW)
(Customer Request matrix 4 Preplan)
(c) Product Characteristics
- Product (Quality) Characteristic
Deployment Chart (QCDC)
- Quality Characteristic
measurements
- Quality Characteristic priority
(DQ-Characteristic matrix 5
House of Quality)
- Bottleneck Analysis
(d) Functions and Mechanisms
- Function Deployment Chart
- Function priority
(DQ - Function matrix 6)
- AHP Function matrix 7
- Mechanism Deployment Chart
- Mechanism priority
(Function-Mechanism matrix 8)
(e) Innovation
- Available or Breakthrough
Technology Deployment Chart
- Technology priority
(Mechanism-Technology
matrix 9)
(f) Parts, Costs and Production Process
- Product Parts Deployment Chart
- Parts priority
(Quality Char.-Parts matrix 10)
- Target Cost of the Parts
- Parts Bottleneck Analysis
- Process Phase priority
(Parts-Process Phase matrix 11)
(g) Reliability
- Fault Tree Deployment
- Fault priority
(DQ-Faults matrix 12)

Fig. 2.11 Framework phase **b**, Customer deployment

Here are instructions on how to proceed.

(a) Before making a *gemba* visit, you first have to identify a specific category of customer and understand how they interact with the product or service. Segmentation and priority of customer type which were defined during the creation of the strategic matrix are very useful for this. See the results of matrix three in the strategic analysis. Some authors (Xie et al. 2003) highlight the need to search for the *future* voice of the customer, beyond the present one. In other words it is important to understand what the future needs will be, not only the current ones, as they arise during interviews with "ordinary" customers. For this purpose we suggest including the so-called "lead user" among the interviewees, i.e. existing customers and potential customers, who can better identify future opportunities, concepts or new products that may arise (Xie et al. 2003).

(b) Before starting interviews a "*State Transition Diagram*" (STD), borrowed from computer science, can help together with QFD, to define decisional processes and different moods experienced by the customer during purchase and utilisation/use of the product/service. The main components of the State Transition Diagram are:

- the state of the system, represented by squares;
- the changing of state, represented by arrows.

Citing Webster's New World Dictionary, a "state" is "a set of circumstances or attributes characterizing a person or thing at a given time; way or form of being; condition" (Yourdon 2010). We can take as an example the diagram that describes the states of a customer deciding to buy a classic bicycle and then while using it (Fig. 2.12).

Some of the states he goes through:

- Idle (stand by);
- Evaluation of the shop;
- Evaluation of the product alternatives;
- Purchase;
- Sitting and still;
- While cycling;
- While braking;
- While parking the bicycle.

Inside this dynamic system, there are changes from state to state, represented by arrows linking pairs of cells. A state can change into different ones; it is also possible the "idle", or "stand by" state, and one or more destination states, beyond which it is not possible to proceed. To complete our diagram, we must finally describe the conditions causing the move from state to state and the actions carried out by the system when state changes; both these aspects are represented next to the arrows.

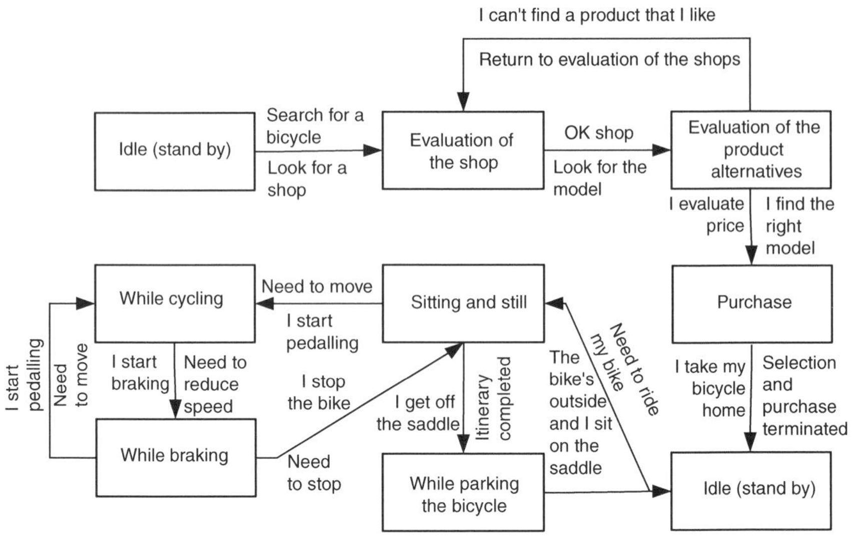

Fig. 2.12 State Transition Diagram for Royal bicycle case study

I suggest the reader starts from a simple diagram, with a limited number of cells; it can change into a more complicated one in your next QFD projects.

We have to remember that State Transition Diagram is created in a subjective way by the team, it does not have an absolute value and it is only a help for the management of information.

(c) How many *gemba* interviews should we conduct? According to Griffin and Hauser (1993) 25–30 interviews for each homogeneous segment of customer are enough to determine 90–95 % of possible product requests. We should clarify that this phase does not aim to determine and organize all the customer requirements, but it seeks qualitative, verbal "raw data" from existing and potential customers. Afterwards the QFD team will use this information base for adding or modifying data in a creative and original way.

According to our experience from 10 to 15 *gemba* interviews are enough to collect a huge quantity of information.

They should preferably be addressed to the customer types indicated in the strategic matrix 3 together with the lead users.

Interviews can be:

- more to ordinary customers, if the type of product does not require high innovation speed or if the risks are difficult for the company to manage;

- more to lead users, if the context is highly competitive, no-innovation no-survival, if the speed of introduction onto the market is a critical point and if the company can take some risks. Information Technology and internet are examples of this second type of market (Xie et al. 2003).

2.5.2 How Is a Gemba Interview Carried Out

The first things you have to do are:

- choose the customer to visit;
- choose the usage context to analyse.

Gemba visits and interviews are normally performed through verbal interaction; interviewers ask open answer questions.

Here are some suggestions for an effective *gemba*.

- A digital camera, a recorder and a piece of paper to write down notes can be very useful: if allowed, conversation should be recorded (this aspect is extremely valuable for reporting all the mood changes and the language used by the customer, once back in the office), taking care that all norms regarding privacy are respected. I suggest taking several pictures of the place where the interview is held.
- While introducing the *gemba* interview to the customer, it is better to explain the purpose of it, i.e. "we need your help to improve our ability to satisfy customer needs". His help is essential, for developing an excellent product, and he must be aware of it.
- We have to understand who we are talking to: a person open to dialogue, a very silent, shy person, a passionate, reflective or instinctive person; and depending on this we must adopt a specific interview style. Being receptive to the information supplied is fundamental, showing empathy to problems highlighted and avoiding conflictual or defensive behaviour. *Gemba* is not selling.
- We can ask open questions to stimulate conversation, such as: "How long and why have you used this product?"; "Can you please show me how you usually use it?"; "What do you like best about the existing products?"; "What don't you like?"; "What do you consider when buying the product?"; "What would you improve in the product?". Questions can be similar to states of the State Transition Diagram.
- We should directly ask to the interviewee to find possible solutions to the problems he mentioned; negative observations are very important, they are incentives for new ideas.
- The *gemba* interview can be rather long (often more than 1 h). When the customer starts repeating the same concepts it is probably time to end the interview.
- Our interlocutor should be asked to show us how he uses the product/service. What is his own State Transition Diagram like?
- The emotional aspects of the customer are very important, which is why a digital camera, a voice recorder, as well as written notes are often essential.

Exercise Firstly ask yourself or top management (if possible) if you can make a written list of customer requests and expectations. Is this information widely available within your company?

Draw a State Transition Diagram showing the states your customer lives through during his product or service experience. Draw a simple STD, with maximum 5–6 states.

Gemba: choose the customer you want to visit and where. Start with few interviews, five for example. Follow the suggestions presented in sub paragraph 2.5.2. Take your camera, a recorder if possible as well as pen and paper to make notes. Remember to note what the surrounding environment, the weather, and also the people are like. How do you feel after the first interview? Were you satisfied with the raw data collected or was the customer not responsive?

How long were the interviews on average?

2.5.3 Voice of the Customer Table

The QFD team has just gathered a sequence of raw data, also called "customer verbations", made up of non filtered sentences, expressed by different interlocutors.

The sentences are contained in the written notes from *gemba* interviews or deduced from the recording of the interview. They represent the "Voice of the Customer" and are placed into the "*Voice of The Customer Table*" (VOCT) (Mazur 1996), a chart that can be set up in different ways, for example according to a 5W1H model, generally made up of distinct sections for every client with the relative context of use (Table 2.1).

The need for the working group to be continuously aware of the predetermined targets, available resources and last but not least the time limit for the QFD research, can sometimes lead to simplification of the 5W1H model above, reducing the VOCT to a list of received sentences and a short introduction of the context for each interview. In any case, the 5W1H model is advisable.

My suggestion is not to screen the sentences; if there are lexical mistakes, common sayings or strong expressions, it is best to leave them, in order to express, as far as possible, the emotion and passion behind the assertion.

Now let's see an example of a simplified VOCT for our project. In the "Royal" case some specialized retailers and adult end users were interviewed. They were selected in areas where the "Royal" brand receives widespread appreciation and traditional bicycles constitute a good market share. Ten interviews were conducted. The list below shows raw sentences from the first interview. You will find the complete VOCT in the appendix to this chapter.

Table 2.1 Royal QFD project Voice of the Customer Table example, according to a 5W1H model

Who is the interviewee	Where interview is conducted	When interview is conducted	What customer notes	Why customer says this sentence	How the problem could be solved
Interviewee 1 shop owner	Bicycle shop	Spring, morning, sunny day	The brakes are disgusting. Do you buy them in the supermarket?	A lot of negative feedback from customers. Interviewee is talking during bike tuning	Improve the overall quality. Spend more on brake pads
			A lot of clients come back indignant just a few hours after purchase and want the brake pads to be changed; a problem of this sort in such an expensive bike is inconceivable!		Change rubber composition of the brake pads

Interview 1

Retailer, specialised shop in the city centre, he sells mainly mountain bikes. Shop window displays and shop give a very good impression.

In the past he sold some of the Royal range.

Raw data.

- The brakes are disgusting. Do you buy them in the supermarket?
- A lot of clients come back indignant just a few hours after purchase and want the brake pads to be changed; a problem of this sort in such an expensive bike is inconceivable!
- The hubs should be of an older style: chrome steel with the Royal brand mark.
- The wheels aren't aligned, the spokes bust after some weeks.
- The shape of the handlebars is very good; don't make any changes to that.
- The dynamo is just a plastic toy it must be replaced completely.
- The mudguards are mediocre.
- The finishing must be improved to the tiniest detail. Its the attention to those tiny details that helps convince a customer to buy.
- This bike can't be the same as the others.
- Make sure the serial number is clear. Now, sometimes there is a skimpy transfer, sometimes not.
- In the past this make of bike was treated like a Ferrari.

Let summarise where we have got to.

1. *Gemba* process consists of a series of open answer interviews with current customers and potential ones. A small number of interviews are possible, 10–12, 20 at maximum. *Gemba* is the place where the product or service acquires its value for the customer, where they are actually used.
2. Choose customers involved in interviews. They can be ordinary customers and/or lead users and/or potential customers.
3. A State Transition Diagram can help to focus the interview.
4. The interviewee must not be influenced by the interviewer's behaviour. The atmosphere must be friendly, relaxed and informal. We absolutely need her/his help. Before the *gemba* visit, keep some questions ready which could be useful when interviewing *those* who are not very loquacious. Do not filter the sentences. Try to express, as far as possible, the emotion and passion of the assertion. Recording the interview is recommended.
5. The customer verbations (raw data) are used to create the Voice of The Customer Table (VOCT), which can have different layouts. Choose simple, basic, clear schemes. Check them with your team.

Exercise First read the sentences from the Royal case study, in the Appendix to this chapter. In this way you can have an idea of how to create your own VOCT. Draw a simple VOCT from your *gemba* interviews with a spreadsheet or text editor. It will be used in the next step. Record the time spent on each phase for future comparison.

2.5.4 Jiro Kawakita KJ Method and the Demanded Quality Deployment Chart (DQDC)

The KJ method was invented by Jiro Kawakita and it is "a system of integrating qualitative data", the most important principle of which is that "data speak for themselves without any preconceived theory or hypothesis". The KJ was initially developed to solve environmental problems and to enable "a holistic presentation of whole circumstances concerned surrounding an environmental problem" (Kawakita 1981).

Jiro Kawakita, born in 1920, led a full life of geographic exploration and studying the environment (he received the 1984 Ramon Magsaysay Award for International Understanding for his work in the Sikha valley in Nepal). He showed a strong inclination to travel from a very young age (visiting the Mariana, Marshall, and Caroline islands).

In the mid 1940s, he became professor of Geography. In 1953 he was the scientific member of the expedition that set out to climb Manaslu—8,163 m (unconquered until 1956), in Nepal. From that year on and for the rest of his working life, Nepal exercised a strong attraction on him. After exploring the region of Dhaulagiri, he then explored the valley in the southwest of Annapurna, which he called the Sikha Valley, in 1963, and began studying life in the villages.

Kawakita understood that the Sikha Valley was suffering from an ecological disaster, caused by progressive deforestation in the villagers' search for new grazing lands which was leading to landslips and soil erosion. Kawakita looked for solutions, inventing a simple method, called KJ Method, which would later become, as we shall see, one of the key elements in the development of a good Quality Function Deployment process.

Starting in 1963 he collected huge numbers of exclusively qualitative elements, through interviews and observations in these mountain communities. The quantitative elements were too complicated for Kawakita to organise. The problem was how to schematise and draw conclusions from a high quantity of subjective and objective elements. Kawakita began writing a specific sentence to describe each aspect, each one on a small piece of paper. Then he went on by grouping sentences according to existing connections or to similarity. The enormous database decreased in size to only ten aspects and two solutions became apparent: cables and cableways to make the transport of wood and material from mountain to valleys easier, and pipes to bring drinking water to the villages.

Although Kawakita was quite sure that the two innovations were a real solution to meet the needs of the Sikha Valley inhabitants, in order to reach their consensus, rather than use a questionnaire he preferred to spend night after night holding numerous, long meetings around the fire in the villages. He later called this process "Key Problem Approach".

Over the following years Kawakita searched for funds to carry out the work. He published *Hasso-ho* book (A method for idea generation), founded ATCHA (Association for Technical Cooperation to the Himalayan Areas), a no-profit and volunteer organisation, focused on the study of the needs of the Himalayan people. In 1982 founded the Japan Creativity Society, whose objectives are to foster and lead research in creativity and in education.

> Slim and erect, Jiro Kawakita carries himself with immense dignity, whether clad in pullover and parka on horseback in the Himalayas or wearing the impeccably tailored suit of a Japanese businessman. His mien reflects his life's work of creating harmony from chaos (The 1984 Ramon Magsaysay Award for International Understanding 1984).

Jiro Kawakita passed away in 1990 at the age of 89.

KJ Method Phases

Let's come back to the QFD process.

What is our objective in this phase?

We have to identify the requirements, requests, desires and expectations of the client starting from the rough *gemba* elements.

That is we have to re-write the sentences collected during the interviews, according to some rules.

These new sentences, the customer's requirements, are also called *Demanded Qualities* (DQs, Akao 1990). They will later be the foundation on which we'll build a structured questionnaire, this time with closed answers, to be submitted to existing and potential customers, with the aim of finding numerical evaluations.

And which tool is more powerful than Jiro Kawakita's in identifying DQs?

In the KJ method some meetings (work sessions) are necessary, because the reworking of rough sentences resulting from *gemba* is often difficult. Into my experience a good rule is not to exceed more than a total of 3 h of work per session. The number of participants can be variable, generally from 4 to 8, coming from different departments and then with different specialisations (marketing, sales, technical office, production, quality). It is advisable, if possible, to involve a lead-user. If the group is limited, the KJ method must follow anyway, in order to maximise both the analytic and creative components.

It is important to elect a supervisor who coordinates development over time of the KJ analysis and checks that tasks are carried out correctly. The supervisor does not carry more weight than any of the other members of the group, during the evaluation of the sentences identified using KJ.

Let's look at the different phases of the process.

1. *Warm-up*. This is an out-and-out preliminary "warm-up". After the supervisor has explained to the QFD team what the target objectives are, each KJ participant, in turn, expresses a short reflection related to the work that the group is going to achieve, for example the problems that could come up or the phases of this process. A copy of the VOCT (the Voice of the Customer Table) is handed out to each participant and all the additional documents are made available: market analyses, customer satisfaction surveys, renderings, drawings, sketches of the product under development.
2. *Distribution* of the sticky notes and individual note-writing. After the warm-up, from 20 to 40 sticky notes or white paper cards are distributed to every member of the brainstorming group. They should then work individually in silence, without interacting with other people, writing a sentence on each note and on each note there should only be one idea, i.e. on every card there will be a sentence that tries to describe a demanded quality, deduced from the raw data. Some good rules exist regarding the writing of this kind of sticky notes. We'll see these in detail shortly.
3. *Collection*, reading and revision of the sticky notes. After a period of time, considered sufficient by the team and adjusted if necessary should there be any delay while writing out sentences, the sticky notes are collected and each one read out to the group by its author. The individual members of the group then rewrite, revise and improve the sentences, following two main objectives:

   - "Abstraction" from the physical properties and features of the product. See Chap. 1, paragraph 1.8 (Creativity and QFD).

- The complete comprehension by the group of what the sentence means. Furthermore, the choices should be made following the principle of "sharing and mediation", not of majority. No votes but agreements are reached.

4. *Deployment* of the sticky notes. We refer to the word "Ten" as used in QFD, the deployment of the kimono borders. Sticky notes are randomly spread across a table or on a large sheet of paper. I suggest one or more white sheets, where you can place sticky notes, instead of the table itself, because it is likely that there will be more than one session. It is important to keep the work done, in a different place from where the last session was conducted.
   The fundamental property of sticky notes comes into play: they can be fixed and detached easily. Sticking them on a large sheet that can be rolled when necessary is a good strategy.
5. The sticky notes are then *grouped* according to subjective criteria (but completely agreed on within the QFD team) of affinity and similarity, in other words moved around on a flat surface (wall, table, board), and organised into categories of up to three levels of detail at the most (often there are no more than two levels).

If, during phase 2, creativity was maximised by silence and the lack of influence between the various members of the group, here it comes directly from playing around with the repositioning and creation of new sentences or new sticky notes, to cover any possible spaces in this deployment.

There is maximum freedom allowed in the creation or deletion of the sentences.

Clearly, sticky notes with similar or identical meanings can be merged into only one sentence.

In this context parent notes are defined. They are the headings that summarise the parameter of affinity used in the specific group. A hierarchical tree of two or three levels is created inch by inch. This structure is called Voice of the Customer Table (VOCT) (Akao 1990).

Some in-depth analyses are necessary.

First of all we can make a list of *good rules to follow* when we build the descriptive sentences of DQs.

(a) In writing sentences the guideline is "abstraction". Every reference to characteristics, performance or components of the product has to be removed or, rather, transformed into client expectation. As a help, the members of the group can ask themselves while reviewing the note that they want to abstract more: "Which expectations, or which customer requests are satisfied by this particular characteristic or ability that the product or service has?"
   Consider for example the raw sentence deduced from *gemba* regarding the bike: "Brake pads are barely efficient for stopping and they screech intolerably". It could naturally produce a reviewed sentence of this kind: "Brake pads less noisy" or "Silent brake pads". However in both phrases reference to the product is still present. The principle of the abstraction must be applied in order to

achieve a good DQ: “Silence while slowing down”. We can see here that State Transition Diagram (STD) can be used to help. It expresses states, our interlocutor finds him/herself in when she/he uses the product. The STD allows us to concentrate on the customer, not on the product.

Within the sentence any links to the physicality of the product, its characteristics and performance, are a recurring mistake for a QFD team dealing with KJ brainstorming at the first time.

(b) Abstraction doesn’t have to be confused with *generality*: namely we can hold on to abstraction and yet reach high levels of detail. Or, rather, we can decide, depending on the complexity of analysis that we want to reach, at what level of detail we will stop the DQ sentence. For example, “feeling of comfort”, deduced from the VOCT, is an abstract and generic DQ; we can use this sentence or expand the “feeling of comfort” into abstract and detailed DQ. The State Transition Diagram is particularly useful here too. The sentence can become: “feeling of comfort getting onto the bicycle”,“feeling of comfort while pedalling”,“feeling of comfort when braking”,“feeling of comfort when braking in the rain”,“feeling of comfort when braking in the rain on a gravel road”,“feeling of comfort while purchasing the vehicle”, and so on.

It is clear that it isn’t necessary or obligatory to insert expanded DQs in the different states of the STD, because the number of these would increase greatly. Certainly the STD allows the QFD team to concentrate, if they agree it is opportune, on a more detailed evaluation of some DQs. Always remember that QFD is a way to help develop original products and services, because these phases are also unique, subjective and original.

(c) How many generic DQs need to be identified? The hierarchical structure with two or three levels isn’t generally symmetrical. As we said above, areas that the team subjectively considers more critical, are richer in DQs. Furthermore some isolated DQs exist that cannot be put into any of the groups. These are the so-called “lone wolves”.

As we’ll see in the next chapter, where there is a certain number “n” of DQs a consecutive matrix generally corresponds with approximately “n^2” cells to analyse. For example 10 DQs often lead to a matrix with at least 100 cells. On the other hand, however, a sufficiently trained working group, with an expert supervisor, can manage structures with a high number of DQs. Some of our projects (see Chap. 5) involved analyses of more than 150 DQ and calculation matrices with more than 6,000 cells. This high number of demanded qualities perhaps clashes with the Jiro Kawakita philosophy that is focused on grouping and selection.

Considering the number of DQs, these certainly have to be correlated to the level of available resources for the QFD project and to the time that can be spent on the brainstorming sessions. Based on my experience, the continuity of the project must be preferred above all; so, rather than deploy a huge Demanded Quality chart, a limited number of DQ is more advisable and, at the same time, to maintain energy reserves so as to continue with the analysis in the following matrices and at least get to the characteristics of the product. In cases where you are working on your first or a usual QFD project, a suggested number of the most detailed DQs lies between 10 and 20.

(d) A good beginning for a DQ can be: "feeling of…", "expectation that", "sensation of ", "need to/of", but the DQs do not necessarily have to start this way.

(e) Comparable DQs have to be converted into absolute DQs and negative DQs into positive DQs. For example, "Expectation of greater comfort being seated" is a sentence that has to move to absolute: "Expectation of high comfort being seated". Also: "Not noisy brakes" has to change doubly into an abstract and positive meaning: "Silent braking".

During several marketing courses I have put forward the same case with the same *gemba* list to the students, grouped in many QFD teams (see Chap. 1). The result was that the KJ helped to write a list of needs, some turned out to be similar among the different groups, others, those not immediately or easily deducible from reading the raw data, turned out completely different and original. This for me was decisive proof that QFD leads towards the development of excellent products but which are different from each other even if the initial input is the same.

It is now interesting to mention the Noriaki Kano model (Kano et al. 1984) very briefly, which tries to classify customer needs, based on her/his reactions when she/he interacts with the product or service (Table 2.2).

- First of all, there are some product qualities that must inevitably be present in the product. If not, they lead to the potential client being displeased: he probably won't buy the product or service; they are the so-called "Must be qualities". On the contrary, if the must-be qualities are already present, the client is indifferent. One example from the automotive industry is the presence of wheels or nowadays ABS or airbag.
- On the other hand, some requests are satisfied by "attractive qualities". If present they leave a positive impression on the buyer, but if not, the buyer is indifferent. An example is GPS offered as standard in a vehicle.
- One dimensional qualities are features of the product expressed by the company or requested by the customer as personalization. If these are present, the client is happy, if they aren't, he isn't. For example the colour of a car delivered to the customer, it is important it should be exactly the same as was ordered.
- "Indifferent qualities" are those characteristics that induce no reaction whether they are supplied in the product or service or not. It is exactly here that Quality

Table 2.2 A recapitulatory matrix of Kano et al. model

		Present	
		☺	😐
Not present	☹	One dimensional quality	Must be quality
	😐	Attractive quality	Indifferent quality

Function Deployment process proves its incredible efficacy. The careful analysis of the process of development of the product, starting from the detailed study of the customer, allows for a significant decrease in the probability that characteristics and functionalities for a product are developed which the customer is totally indifferent to.

Note that Kano's model refers to product characteristics. It would be better to switch to customer requests (abstraction) and ask ourselves what customer needs are satisfied by attractive qualities, for example. It could be a good exercise.

Again, with reference to Kano's model and coming back to the test carried out with students that started a KJ from the same raw data, we could make a couple of observations:

- Demanded Qualities similar for the various groups could be associated with Kano's one-dimensional and must-be quality;
- Creative Demanded Qualities (different for the various groups) could be associated with the Kano's attractive quality.

It is only a hypothesis but it could be the start for future studies.

Let's summarise.

There is an important aspect to remember. Three factors contribute to the development of Demanded Quality:

- Current Voice of the Customer, correlated to ordinary clients;
- Future Voice of the Customer, correlated to lead-users;
- Creativity generated during KJ sessions.

Exercise Do not continue reading while you do this exercise. Hide the list below with a piece of paper. Do not read it as it contains a possible solution to this exercise!

Go to the appendix and use the Royal VOCT as a base for the KJ brainstorming session. Remember what you must not do. The aim for you and your group is to create a "Royal" Demanded Quality Deployment Chart with approximately 10 demanded qualities.

Once you have done this job, carry on reading and compare your findings with the DQ presented in the book. What are the differences? Is there a good level of abstraction in your DQs? Check this last point and try to raise the sentences that seem too close to the product/service performances to a higher level of abstraction. Remember: we are talking about the customer, not about the product.

Record the time spent on this phase for future comparison.

Now, having had a little training with Royal bicycle raw data, use your VOCT to work out about 10 sentences and create your own Demanded Quality Deployment Chart.

Now let's come back to our case study.

We can present an example of Demanded Quality Deployment Chart (DQDC) for the Royal project. This is only a subjective example, not the absolute representation of customer requests. Our DQDC with one level is:

- Feeling of refinement and classic style;
- Saving time in assembling;
- Riding a status symbol;
- Make an investment for the future;
- Absence of back stress;
- Regular progression in braking;
- Feeling of silence while riding;
- Feeling of fluidity while pedalling;
- Need for sports style;
- I feel different from the other cyclists;
- Feeling of craftsmanship.

2.6 Customer Analysis. Questionnaire Design and the Preplan Matrix

When the DQDC is created, and starting exactly from that point, the following step in our QFD project consists of submitting a *questionnaire* with closed answers to a great number of customers or potential ones, to obtain quantitative data.

Today surveys can definitely be conducted more rapidly than in the past. If some years ago, this QFD phase was practically one of the biggest obstacles for the success of a QFD project, because of long lead time and costs, now internet and web based software have speeded up this activity, at the same time reducing costs in such a way that small companies or offices can conduct an incisive, low-cost, fast QFD survey.

Once the numerical data from the QFD questionnaire has been collected, it is used as averages in a matrix called "Customer requests matrix" or "Preplan" or "Quality Plan". Output from this matrix is the Demanded Quality Weight (DQW), which keeps track of the importance of customer expectations, competitors' benchmarking and sales issues.

Let's see what happens.

2.6.1 The Questionnaire

If possible, the interviewee should test the current product and the competitor models in a show room or directly at home. For Royal bicycle we have two competitor products that we'll call Product X and Y.

On the other hand, the questionnaires are often filled in by customers, lead users and resellers, who already know the product and think they know its strengths and weaknesses compared to the competitors. If this should be the case the preliminary test isn't necessary.

Now we examine in depth the questionnaire as a practical tool.

It could seem that it only requires appropriate organisational and management knowledge. Actually, we have to support this knowledge with a long and careful study of the potential and limitations of the questionnaire so as to avoid distortion in the data gathered.

The answers that people give to the questions are always the result of a complex process of communication, comprehension and data processing.

Numerous are the definitions that we can give for the term "questionnaire". In the completely general meaning it is defined as a tool that allows the researcher to obtain information with the help of standardised questions.

A good questionnaire, according to Sheatsley (1983), must:

- fulfil the purposes of the research;
- gather the information in the most careful way possible;
- do this within the limits of available resources.

Planning of a Questionnaire

Planning is achieved in various moments (Sheatsley 1983):

1. definition of the generic content areas that must be inserted in the questionnaire;
2. definition of the detailed contents and formulation of the questions;
3. selection of the order of questions;
4. check by using a pilot questionnaire.

As we will see in the next pages, the first and the second points are well-defined in the QFD process.

However, as regards the order of the questions, there are no fixed rules; rather it is decided by an optimal blend of the researcher's experience of and some criteria, which we can find in the specific literature regarding questionnaires (Rattazzi 1990):

- grouping questions that concern the same subject;
- creating uniform groups of items, if necessary preceded by short instructions about how to answer the question;
- general answers about a topic should precede specific ones;

- complex answers that need particular attention should be located in an intermediate position: some studies have shown that generally attention and interest grow at the beginning of the questionnaire, reach a maximum at which they level off, then decrease;
- if you use a temporal sequence, it is better to place questions about the past first;
- Goode and Hatt (1952) remember that the start of a questionnaire is a very delicate moment, because this is the point when real collaboration with the person interviewed begins. So the use of interesting and non aggressive questions at the beginning is suggested.
- the most aggressive questions or those that reveal the fundamental purpose of the research should come at the end of the questionnaire;
- if possible, use questions whose function is to check the reliability of some answers;
- avoid facilitating "response set" circumstances. This means that sometimes there is a tendency to always answer the questions in the questionnaire in the same way, whatever the question content or the correct answer may actually be (Bailey 1985).

To summarise, we could say that the rational order of the questions must be integrated with an order that takes into account any possible psychological reactions on the part of the interviewee (Pellicciari and Tinti 1987).

As concerns the fourth point, normally QFD skips the pilot survey with a prototype questionnaire.

Practical Tips and Hints

Before we analyse methods for formulating the questions and the problem of the so-called "response effects", it is worth touching on some practical tips and hints on how to improve the questionnaire.

In addition to paying careful attention as to how the questions are expressed and interpreted, they must also be written in suitable, understandable language, so that you avoid falling into the "*expert trap*". Also, graphics and printing must also be planned in detail so as to contribute, sometimes in a decisive way, to a "friendly" look and to avoid mistakes.

The *length* of each question and of the questionnaire itself is important; the rule, generally recommended by the literature, suggests limiting the length of questions as much as possible. In fact you should avoid influencing the interviewees' answers. Questionnaires must not be too long. If they are, various effects that decrease the quality of the data gathered occur: a reduction in concentration, casual answers, the "straight line responding" effect which appears in some completed questionnaires as the tendency to use the same answer from a certain point of the questionnaire and thereafter.

Even if it may only seem a problem that concerns style, experience suggests great care must be taken when choosing the *title* of the questionnaire. Although some authors suggest the purposes of the research should not be mentioned explicitly, in order to avoid answering strategies different to those that would adopted under real life conditions (Rattazzi 1990), I disagree with this, when it comes to a QFD questionnaire, as in this case, the interviewer *must explain the aims of the survey with complete transparency.*

We need he customer's help to develop exceptional products or services.

We need to understand her/his opinion about us, our strong points and our weaknesses. She/he will develop the product or the service with us, on our side, not on the opposite one. We need to create a collaborative chain between customer and us with rules based on mutual trust. The starting point is transparency.

After the headline, instructions and rules about how to complete the questionnaire must be stated clearly.

Questions

In a generic questionnaire the questions can be of two kinds:

1. open questions (or with free answer);
2. closed questions (or with fixed alternatives).

In the first case the answers are free; they can be limited only by the space available on the form. They are useful when it is impossible to predict the whole list of alternatives or also when, if known, they are too numerous. The main limitation for these appears to come during the phase of sentence management, because they often contain useless, irrelevant, unclear or badly expressed information.

In closed questions the interviewee is invited to answer using a list of alternatives; this list can help to understand the question better and to focalise just on those aspects that the researcher has in mind. Anyway, closed questions also present various problems and preparing them is generally an arduous task. Among other things, we need to cover an ample range of individual reactions with a very low number of distinct and independent sentences.

It is impossible to say which typology, open or closed, is better. On the one hand, closed questions permit quantitative data analysis (this is the case for QFD); on the other hand *gemba* and the implementation of KJ, based on sharing, and dialogue, are powerful research tools.

Answers

Answers for closed questions can be divided into (Pellicciari and Tinti 1987):

1. answers with forced choice;
2. graduated answers;
3. answers with a preference list.

In the first case the subject has two opposing choices, for example "yes" or a "no", "agree" or "disagree".

In questions with graduated answers it is possible to choose from various categories of answers, presented in a particular order (sometimes they constitute a "rating scale").

In answer to a question with a preference list, the interviewee makes a choice from a list of possible answers.

QFD questionnaires require the so-called "rating scales". In this case the interviewee gives an assessment, which she/he can choose from a well-defined range of answers, for every alternative. These techniques are called "non representational measurement". We try to derive behaviour and valuations, asking the interviewee to express them using a categorical or numeric assessment (Dawes and Smith 1985). The interviewee must choose only one category, mark a number or make a cross on a line. A typical example is the so called "Likert scale". Selecting the verbal expressions, researchers make two hypotheses, which allow them to attribute some weights to every sentence:

- the same distance of "meaning" between one answer and the next;
- the interviewees interpret the answers in the same way.

Response Effects

Specific errors exist in the completion of a questionnaire due to the formal features of the question or the personality of the interviewee. These errors are called "*response effects*" (Rattazzi 1990).

Questions can include some elements both in the text of the question itself, as well as in the list of answers, that can create distortion of various types. Though formally well balanced, questions could include expressions or words that could influence the interviewee's answers.

In a questionnaire, questions are never isolated but part of a group. The context and position of every question affects answers. The situation is particularly serious when the data are used to study behaviour or opinions. Some coherence effects can occur when subjects try to show that they are being coherent in their answers, i.e. by answering a specific question not honestly, but referring to their behaviour for another question (Rattazzi 1990; Trentini 1980).

There is the risk that the answers, selected by the interviewee, cause accessibility to some information, that could be used to answer successive questions.

Distortion is possible as an effect of the structure of the alternative answers.

Here are some examples of good practice.

- Random assignment of questions and alternatives to answers, a common feature in web assisted surveys today.
- Increase the interviewee's motivation, explaining that some questions will be difficult and that concentration is necessary.
- Simplify the interviewee's task, shortening the list of alternatives, where it is possible.

To conclude, it can be said that what is important to remember is not how the answers will be influenced by variables, but only that they will be, and that we have to find all strategies to reduce the response effects as far as is possible.

QFD Questionnaire

Let's go back to our QFD project. Two key QFD questions will be placed in the centre of the questionnaire. The beginning and the end include free topics: for example sales or technical research, or customer satisfaction surveys. In other words, only two questions are binding for the QFD framework and, most of the time, only these two questions form the entire questionnaire.

(1) The first question concerns *an evaluation of the customer's propensity to buy.* Namely it asks *how important it is to satisfy a customer request (DQ) in order to persuade the interviewee to buy that type of product* (in the Royal case, a classic bicycle).
In this question we do not evaluate brand diversity, we do not ask for an opinion about a specific product or model but only the level of importance.
The DQ involved in the questionnaire are those which have a higher level of detail in the DQDC.
In the case of questionnaires designed for services, the question becomes how important is it to satisfy a customer request (DQ) in order to persuade the interviewee to make use of that type of service.
Answers are ranked on a 5 point Likert scale, with the following associated statements:

1 = not important at all
2 = a little important
3 = rather important
4 = very important
5 = extremely important.

(2) The second question *concerns the assessment the interviewee gives to the ability that products/services (ours and competitors), involved in the benchmarking, have to satisfy the Demanded Quality in question.*

Again, answers are ranked on a 5 point Likert scale, with the following associated statements:

1 = very bad
2 = insufficient
3 = sufficient
4 = good
5 = very good.

In this case, every DQ will have an assessment for both our existing product/service and those of our competitors. Thus the benchmarking process has been done.

To summarise, an example scheme can be as follows.

- Survey Title.
- First Area. Friendly questions, hobbies, interests, basic identification of the interviewee: male/female, avoid questions about studies, type of work, employed/unemployed, family status, if necessary these questions that can posed at the end of the questionnaire.
- QFD Area.

(a) How important is it that these customer requests are satisfied in order to persuade you to buy a classic bicycle? 1 = not important at all; 2 = a little important; 3 = rather important; 4 = very important; 5 = extremely important). Please select only one answer per row.

Demanded quality 1	1..2..3..4..5
Demanded quality 2	1..2..3..4..5
Demanded quality 3	1..2..3..4..5
....	....
Demanded quality n	1..2..3..4..5

(b) Please give an assessment of the ability that these products/services (ours and competitors) have to satisfy these customer requests (1 = very bad; 2 = insufficient; 3 = sufficient; 4 = good; 5 = very good). Please select only one answer per row and column.

	Royal	Product A	Product B
Demanded quality 1	1..2..3..4..5	1..2..3..4..5	1..2..3..4..5
Demanded quality 2	1..2..3..4..5	1..2..3..4..5	1..2..3..4..5
Demanded quality 3	1..2..3..4..5	1..2..3..4..5	1..2..3..4..5
Demanded quality n	1..2..3..4..5	1..2..3..4..5	1..2..3..4..5

- Conclusive Area, commercial or personal information. Close by thanking the interviewee for his help in developing excellent products or services.

Questionnaire distribution can follow traditional modes, with the use of sales network, for example, to improve the percentage of returns, or fax or postal mailing. In my personal experience, questionnaire surveys and data management have been the main bottleneck in a QFD project, because of the time spent (months) and the costs.

But today internet allows for exceptional performance at definitely lower costs than classic techniques and at a speed that cannot be compared to the past.

Web sites like www.surveymonkey.com, www.survio.com, www.freeonlinesurveys.com, www.kwiksurveys.com, www.sogosurvey.com and others, allow a questionnaire design to be sent via remote server to a mailing list, normally extracted from our email client reader.

Data gathered are processed according to the website's user specifications. We can set up filters, e.g. geographical or depending on the interviewee's profile, as well as personalised reports. We used this kind of approach in one of the case studies presented in chapter five, with exceptional results, from the point of view of speed, cost and data available.

Data can generally be exported into spreadsheets (.xls or .csv format).

Online surveys are destined to revolutionise this phase, making them definitely more accessible (high speed and low cost) for small companies to use in a Quality Function Deployment project (see Chap. 5 case study 1).

Exercise It is time to create your questionnaire, starting from the Demanded Quality Deployment Chart relative to your project.

Plan a simple survey and use one of the online survey tools, available on internet. Collect email addresses from business partners, existing customers and potential ones, friends and lead users. Then, use these email addresses as the database for your survey. Select the online survey website that seems to better suit your needs and/or you like more. Set your budget (at the present, for example, if you can keep it to less than 1,000 respondents then the cost is very low)

Design your QFD questionnaire online and submit it (a good starting panel can be about one thousand contacts). Wait for the answers to be collected and then have fun using the customised reports.

2.6.2 Preplan Matrix and Demanded Quality Weight

You know that information from questionnaires collected physically or online, gives numerical data, which can be processed using the common statistical techniques.

What kind of information do these data bring with them?

For every Demanded Quality, some numbers, between 1 and 5 (Likert scale) are available:

- average buying propensity (from first question);
- average competitors' benchmarking.

It is now possible to insert these values into the Customer requests matrix, which Akao called Preplan (Akao 1990). The Customer requests matrix No. 4 aims to

	CUSTOMER REQUESTS MATRIX 4 (PREPLAN)								
		BENCHMARKING			PLAN			WEIGHT	
CUSTOMER REQUEST (DEMANDED QUALITY) DEPLOYMENT CHART	DEGREE OF IMPORTANCE	Royal	Competitor A	Competitor B	ASSESSMENT TARGET (Quality Plan)	RATE OF IMPROVEMENT	SALES POINT (⊙=1.5 ⊘=1.2)	ABSOLUTE WEIGHT	CUSTOMER REQUEST PRIORITY DQW (%)
Feeling of refinement and classic style	4.8	4.0	4.0	4.0	4.5	↗ 1.13	⊙	8.10	***11.16%***
Saving time in assembling	4.0	2.5	3.0	3.5	4.0	↗ 1.60		6.40	***8.82%***
Riding a status symbol	4.8	4.7	4.0	4.4	4.7	→ 1.00	⊙	7.20	***9.92%***
Make an investment for the future	4.7	3.1	4.7	3.3	5.0	↗ 1.61	⊘	9.10	***12.54%***
Absence of back stress	3.2	4.1	4.3	3.5	4.8	↗ 1.17		3.75	***5.16%***
Regular progression in braking	4.6	3.1	4.5	3.7	5.0	↗ 1.61	⊘	8.90	***12.27%***
Feeling of silence while riding	3.8	2.7	3.0	3.6	4.1	↗ 1.52		5.77	***7.95%***
Feeling of fluidity while pedalling	3.5	3.2	4.0	3.0	4.5	↗ 1.41		4.92	***6.78%***
Need for sports style	2.0	2.0	3.0	2.7	3.5	↗ 1.75		3.50	***4.82%***
I feel different from the other cyclists	4.1	2.5	3.0	2.5	3.5	↗ 1.40	⊙	8.61	***11.87%***
Feeling of craftsmanship	4.8	2.3	3.0	3.0	3.5	↗ 0.88	⊙	6.30	***8.68%***
							Total	72.55	100%

Fig. 2.13 Royal Customer Requests matrix 4 (Preplan)

quantify customer requests with a weight. As we'll see in the next pages, this index summarises:

- average buying propensity;
- average competitors' benchmarking;
- brand image and sales strengths.

Now we can analyse a typical Preplan matrix scheme (Fig. 2.13). You may find the discussion in the following pages somewhat complex and therefore we suggest you make frequent reference to Fig. 2.17

(a) On the left of the matrix you can see the structure of the Demanded Quality Deployment Chart (DQDC), in this case created with only one level of detail. The average propensity of the customer to buy is placed n the first column on the right of the DQDC, called *Degree of Importance* (DI). It is the average of the numerical answers to the first QFD question in the questionnaire.

(b) The three following columns relate to average assessments from interviewees about products or services (answers to the second QFD question); we can call these answers "*benchmarking*" or "competitive analysis".
When there is only one competitor, there are two columns, our company and the competitor. If we are not able to get information about competitors, there will be only one column, that of our company.

(c) The assessment target, also called "*Quality Plan*", on the right of the benchmarking area, is the target of perceived quality that the development team calculates for every Demanded Quality (DQ), starting from the opinion given about our product and about those of our competitors. The logic that this area of the matrix follows is:

- if our product has a higher average score, an assessment improvement to the specific DQ is unnecessary;
- if our product has a lower or equal average score, an assessment improvement to the specific DQ is necessary. The assessment target in this case increases and assumes a value greater than the best score for our competitors (at most 1 unit more) but not more than 5, which is the maximum limit for the Likert assessment scale. Should the assessment values for both our company and the competitor be the same, then it is generally preferable to increase the Quality Plan. From my experience it is best to always apply an increase of 0.5 points with reference to the best score.
  Let's consider, for example, Demanded Quality "Feeling of refinement and classic style". Average scores are 4.0 ex aequo, Quality Plan is then:
  "Feeling of refinement and classic style" Quality Plan = 4.0 + 0.5 = 4.5
  For Demanded Quality "To make an investment for the future", our Royal bicycle got an assessment which was certainly lower than the first competitor and its Quality Plan rank is with a score of 5.0.

(d) For every DQ, the *ratio* between Quality Plan and average assessment on our product (first column of the benchmarking area) is called "*Rate of Improvement*" (RI). Rate of Improvement evaluates the effort we have to make in order to reach Quality Plan from the present assessment.
RI is:

- equal to 1, if we are market leaders for that DQ according to the average customer assessment;
- more than 1, if there is a gap between present assessment and target Quality Plan.
  Akao adds an arrow to this calculated value, which only has the function of indicating if we are leader or follower/equal in the benchmarking.
  The arrow is:

- horizontal, if that DQ is leader in assessment (RI = 1);
- sloping, if that DQ is follower/equal in assessment (RI > 1).
  For example, "Regular progression in braking" obtains a rate of improvement of 1.61, among the highest, as a result of Quality Plan (5) divided by average assessment on Royal bicycle (3.1).

(e) The next column represents the so-called *Sales Points* (SP). It shows graphical symbols, associated to the Demanded Qualities. A Demanded Quality is a sales point if it represents a key point for the company, from the point of view of brand image communication or if it is a good argument for sales.

Sales Points are generally identified with the help of top management.
There are three possible symbols:

- ⊙ = strong sales point; score 1.5;
- ⦸ = medium sales point; score 1.2;
- no symbol: this DQ is not considered a sales point; score.

Again the graphic features of Quality Function Deployment appear, strongly oriented to the use of symbols and to a universal, direct language.
It is a good rule to select few sales points, with a column filling not more than 30–40 %.
In the Royal bicycle case study strong sales points are: "Feeling of refinement and classic style", "Riding a status-symbol", "I feel different from the other cyclists", "Feeling of craftsmanship"

Exercise Before reading further, read the meaning of the columns of the Preplan matrix up to this point again. Now ask yourself how you could trace the information displayed for every DQ and collapse it into a single numerical index. In other words try to find an overall value for the Demanded Qualities.

(f) Now continue reading and compare your index with the so-called "*Absolute Weight*".
Absolute Weight (AW) is the product of Degree of importance (DI), Rate of Improvement (RI) and Sales Points (SP):
Absolute Weight (AW) = DI ∗ RI ∗ SP.
In this way, for each Demanded Quality, the Absolute Weight brings together the average buying propensity of the customer, the gap between our and our competitor's present assessments, the target assessment and the company's commercial sales policy.
This process happens for every customer request.
As you can see, QFD synthesis is complete and robust.
(g) In the end, like in every QFD matrix, Absolute Weight becomes Relative Weight that is its value in percentage. Relative weight is also called Demanded Quality Weight (DQW) or Customer requests priority.

Like strategic matrices (Paragraph 2.4) DQW values con be sorted in order and drawn in a descending graph (Fig. 2.14).

Again, as in the strategic matrices deployment, the DQW trend can summarise which are the most important customer needs to be satisfied and which are less important.

In the Royal case study we see that the first four DQ highlight the need for exclusivity and refinement. The main target is to develop a product that satisfies the need to make a good investment for the future, for reliability and economic value.

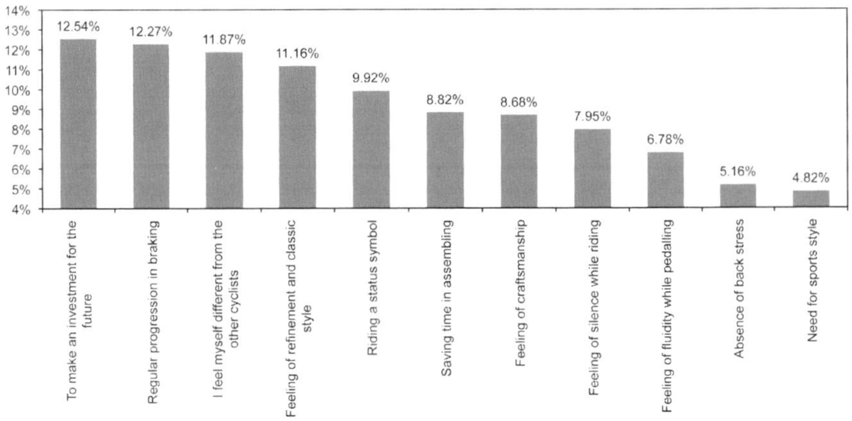

Fig. 2.14 Histogram of the Royal Customer Request priority

A more technical issue is the request for regular progression in braking, that gets a relatively high weight (it is in second place), because of the great gap between the assessment given for our product and that of "competitor A" and because our company wants a decisive improvement for this aspect, to be used as a sales point (medium value).

If we have to describe the direction that the customer wants for Royal, between classic style and sports style, it is clear clients, on average, suggest a move towards the first one. "Need for sports style" comes last in the diagram.

Other similar comments about the results of this matrix are possible.

In practical terms a clear and powerful input for technicians and designers is ready.

We learn from Preplan matrix that Quality Function Deployment is a strong analysis tool that can break down very complex problems, such as the development process for new products or services, into an orderly sequence of elementary sub-problems.

Thus two results are possible:

- simplicity;
- methodical risk reduction of developing the "wrong" product or service because of a system of problem solving which is focused on the product as a whole.

But this is not all.

In fact, as we have become used to doing with QFD, we can use the Demanded Quality Weight as an input for a new matrix. In the next chapter we will leave customer analysis and *move to the technical world*, to the design and test offices, and first to our product technical features deployment.

Exercise Let's create a difficult matrix: the Preplan of your QFD project. Take the Royal case study as a guideline for your project. Where possible, set up the formulas needed to automate computation in the software spreadsheet. Follow these stages:

(a) survey data management and average calculation;
(b) spreadsheet set up with formulas;
(c) insert Demanded Quality Deployment Chart;
(d) average results from question 1;
(e) average benchmarking assessments;
(f) quality plan should be calculated automatically;
(g) sales points selection;
(h) absolute weight and Demanded Quality Weight can be obtained automatically with simple settings of formulas. Take note of the time and energy spent in this phase.

2.7 Appendix. Royal Voice of the Customer Table (VOCT)

Interview 1

Notes. Retailer, specialised shop in the city centre, he sells mainly mountain bikes. Shop window displays and shop give a very good impression.

In the past he sold some of the Royal range

Raw data

The brakes are disgusting. Do you buy them in the supermarket?

A lot of clients come back indignant just a few hours after purchase and want the brake pads to be changed; a problem of this sort in such an expensive bike is inconceivable!

The hubs should be of an older style: chrome steel with the Royal brand mark.

The wheels aren't aligned, the spokes bust after some weeks.

The shape of the handlebars is very good; don't make any changes to that.

The dynamo is just a plastic toy it must be replaced completely.

The mudguards are mediocre.

The finishing must be improved to the tiniest detail. It's the attention to those tiny details that helps convince a customer to buy.

This bike can't be the same as the others.

Make sure the serial number is clear. Now, sometimes there is a skimpy transfer, sometimes not.

In the past this make of bike was treated like a Ferrari.

Interview 2

Notes. Retailer, specialised shop, selling country bikes and mountain bikes, good exhibition of classic bicycles. Shop window displays and shop well presented.

Raw data

He brings us a Royal and speaks to us outside the shop.

In this country the traditional bike represents just 5–10 % of the total. It's a market niche, for classic bike-lovers now.

Certainly MTB are still going strong and most clients are moving to that type.

Who buys Royal? Certainly not young people.

I don't know how much the market is worth; anyway here it can easily overtake 30 % of total sales.

There a "problem youths": the younger generations don't know the brand as well as middle aged or older customers.

Of course the brand gives an air of prestige.

At one time, the owner of your factory used to get on every bike that came off the production line.

If you enhance quality, a new market can be developed. The guy, who buys Royal, thinks he is buying a good product and yet when he gets it home he is disappointed.

We sellers should push more.

It is a brand that gives confidence and quality. New products: a good quality Ladies country bike, a good quality men's sports bike and a review of the quality of the current City bike, because, now, quality in comparison to price is rather poor.

A common request was noticed: to keep the styling of the range: classic and conservative.

Interview 3

Notes. Retailer, shop specialised in classic bikes and mountain bikes, good exhibition of classic bikes. Polished design of the shop.

Raw data

Half of my sales are for classic bikes, you know, bikes for bankers! Of course in the past it was the standard for quality bikes, but now, you can see here, and here and here: plastic, rubbish.

Who buys a Royal? People aged from 30 to 40 upwards, young people don't know the brand.

In my town two big companies give a Royal to people when they retire.

It is a bike that never needs to be set.

Good bikes always get sold.

I could win back custom I have lost over these last few years.

The current Royal is too bare, keep to the traditional style.

Improve reliability.

The bike headset slackens easily, it has manoeuvrability problems. It doesn't work well, we don't know why.
I don't like the quality or the style of the pedals.

Interview 4

Notes. End user, Royal customer. 55 years old, male. Interviewed after a bike ride in the city, about 5 km, sunny day, mild weather.

Raw data

The supermarkets (big retailers) only sell rubbish.
You ask about usage? Who knows, it depends. I would say, mainly in people's free time, but also for work. I would say fifty and fifty. Yes, it is also a way to get around, but here bikes are not used as much as in other countries.
A bicycle with Royal characteristics can be recognised even if doesn't have the brand marked on it.
The brake pads are not very efficient when stopping and they screech intolerably.
Sometimes the pedal doesn't turn if you back pedal.
The wheel rims must be improved.
We, Royal clients, pay very close attention to detail and to the classic style.
Keep and increase the gold finishes.
The Royal bike is the Rolls Royce of bikes.
The Royal bike doesn't have very good accessories; there are few gadgets even if I don't think they are very important.
The dynamo is black. Now just tell me if a black dynamo can be fitted on a bike like this! It should be silver!

Interview 5

Notes. End user, Royal customer. 45 years old, male. Interviewed in a bike shop.

Raw data

It's just a slightly-known brand where I live.
Middle-aged people buy them because they consider them the top of the range in bikes.
People really admire Royal bikes more than any others.
It is a guarantee of quality.
Quality must be kept at a high level.
Use aluminium rims with steel spokes, but keep the classic style.
The handlebar plating, is barely sufficient.
Plastic on the Royal? That's crazy! Only leather.
I still remember the bone grips; of course you can't have those nowadays but…
The mudguards are ok. The gold finishing touches are good but some transfers are not very resistant and not suitable for this bike.

Interview 6

Notes. End user, Royal customer. 57 years old, male. Interviewed before and after a bike ride on a Royal.

Raw data

The saddle is good as it is.

Use brakes with small pistons.

The light system should be reviewed: a bigger light, better finishing, perhaps with the Royal brand.

Better not to speak about the kickstand; the spring is too stiff and makes it feel like a supermarket bike.

Why you don't you use almost black and also the old "Havana" colour ways?

The frame is perfect.

The chain guard is very good.

There is a problem with advertising: you should think of suitable advertising support. You should certainly improve the quality of the product. Take a lesson from the past.

Interview 7

Notes. Retailer, shop specialised in city bikes and mountain bikes, few classic bikes. Interview conducted standing beside a Royal bicycle.

Raw data

The market is suffering from distribution in supermarkets, bikes that "bend" if you just look at them.

It's a brand that has its charm as a guarantee of quality, even if nowadays it leaves much to be desired.

I am indifferent to a re-launch of the brand because, at the present moment, demand for it is rare.

Keep the range as it is, there are also too many models out and about. You already make sport bikes under other brands.

You often use cloth covered brake pads that cuts out noise but makes braking less efficient; use aluminium rims: braking is better even if you lose a bit of the classic effect.

Don't use three part hubs.

Why don't you fit a forged point fork?

Interview 8

Notes. Retailer, shop specialised in city bikes and mountain bikes, moderate variety of classic bikes. Interview conducted standing beside a client Royal bicycle undergoing maintenance.

Raw data

Usually people who buy Royal bikes have money to spend and they think with their own minds.

When people ask for a good bike, they ask explicitly for a Royal or someone recommends a Royal.
The prestige bicycle market has few brands on offer: this is a very tiny niche market. New development could certainly re-launch the brand, but the numbers will be always limited.
The range should be extended.
We need a sports bike, but with a closed chain-guard like in the past.
On pedalling hard the pedal linchpins bent.
The hubs aren't even lubricated!
A black Royal is like a red Ferrari.

Interview 9

Notes. End user, Royal customer. 49 years old, female. Interviewed before and after a bike ride on a Royal.

Raw data

Everybody, both young people and adults, knows this brand.
Nothing else like it exists on the market.
As the range evolves there should be a model which is a little more modern in style, but has more sophisticated technical features.
Why not dream up Royal accessories: a leather or wainut keyring, a basket, a coat protection system, newspaper rack, Royal brand bags, Royal brand padlocks?
Leather brake levers.
There is some noise coming from the pedals.

Interview 10

Notes. Retailer, shop specialised in city bikes and mountain bikes, moderate variety of classic bikes. Interview conducted with standing beside a client Royal bicycle undergoing maintenance.

Raw data

It's a well-known brand.
I am certainly optimistic about a re-launch of the brand.
I would be happy to offer a traditional yet different bicycle.
We should make the sports cycle like a long time ago with or without gears and a closed chain-guard.
The range is too limited both in the number of models and the colours available, add a ladies country bike model.
Even though the saddle is good enough, it should be improved and made more elegant and classic. Some customers ask to change it and fit a "Brooks" saddle.
The luggage rack should be improved.
The Royal was very much appreciated at the trade fair.

References

Yoji Akao (ed) (1990) Quality function deployment QFD integrating customer requirements into product design. Productivity Press, Cambridge

Bailey KD (1985) Metodi della ricerca sociale (Methods for social research). Il Mulino, Bologna

Dawes RM, Smith TL (1985) Attitude and opinion measurement. In: Lindzey G, Aronson E (eds) Handbook of social psychology. Random House, New York

Galunic DC, Rodan S (1998) Resource recombinations in the firm: knowledge structures and the potential for Schumpeterian innovation. Strateg Manag J 19:1193–1201

Gemba Definition. http://en.wikipedia.org/wiki/Gemba. Accessed 3 Apr 2013

Goode WJ, Hatt PK (1952) Methods in social research. McGraw-Hill, New York, NY

Griffin A, Hauser JR (1993) The voice of the customer. Market Sci 12(1):1–27

Kano N, Seraku N, Takahashi N, Tsuji S (1984) Attractive quality and must-be quality. J Soc Qual Control 14(1984):39–48

Kawakita J (1981) The use of a holystic presentation of "Key Problem Approach" in a technical cooperation project for a Himalayan Hill Area. MAB/CONF-81/5/6 UNESCO-ICSU Conference-Exhibit: Ecology in Practice, Paris 22-29 sept. 1981

Masaaki I (1997) Gemba kaizen: a commonsense low-cost approach to management. McGraw-Hill Professional, New York

Mazur G (1996) Comprehensive quality function deployment for products, Implementation guide, Michigan Engineering, University of Michigan, Ann Arbor Michigan

Mazur G (1999) Comprehensive quality function deployment for products. Product version 2000. 1999 QFD Network

Pellicciari G, Tinti G (1987) Tecniche di ricerca sociale (Tecniques for social research). Franco Angeli, Milano

Prahalad CK, Hamel G (1990) The core competence of the corporation. Harvard Bus Rev 68 (3):79–91

Ramon Magsaysay Award for International Understanding (1984) Biography of Kawakita. Manila. http://www.rmaf.org.ph/Awardees/Biography/BiographyKawakitaJir.htm. Accessed 13 Jan 2014

Rattazzi AM (1990) Il Questionario (The questionnaire). CLEUP, Padova

Ramadhan RH, Al-Abdul Wahhab HI, Duffuaa SO (1999) The use of an analytical hierarchy process in pavement maintenance priority ranking. J Qual Maintenance Eng 5(1):25–39. MCB University Press, pp 1355–2511

Ronney E, Olfe P, Mazur G (2000) Gemba research in the Japanese cellular phone market. Transactions from the twelfth symposium on quality function deployment/6th international symposium on QFD, QFD Institute, June 2000, Ann Arbor, MI

Saaty TL (1980) The analytic hierarchy process. McGraw Hill, New York

Sheatsley PB (1983) Questionnaire construction and item writing. In: Rossi PH, Wright JD, Anderson AB (eds) Handbook of survey research. Academic Press, New York

Trentini G (ed) (1980) Manuale del colloquio e dell'intervista (Manual of meetings and interviews). Isedi, Milano

Xie M, Tan KC, Goh TN (2003) Advanced QFD applications. American Society for Quality, Quality Press, Milwaukee

Yourdon E (2010) Just enough structured analysis Chap. 13 state transition diagram. http://yourdon.com/strucanalysis/wiki/index.php?title=Chapter_13. Accessed 7 Oct 2013

QFD from Product Characteristics to Pre-production

3

Abstract

In this chapter the rest of the QFD framework phases will be discussed:

- Product characteristics;
- Functions and mechanisms;
- Innovation;
- Parts, Costs and Production process;
- Reliability.

The reader can continue developing her/his QFD project, using the exercises as a guide. The output from Preplan is the customer requests priority, also called Demanded Quality Weight. This index is used in creating the Quality Function Deployment House of Quality in order to obtain the priority of the product Quality Characteristics. Characteristics help to build the product Functions (for a bicycle, "It lights up the road ahead" or "It brakes"). Functions are weighted from both the customer and the technician's point of view. In the QFD model, Functions then lead the team to the Mechanisms, macro-systems which carry out the product functions, for example, the "Steering mechanism" or the "Transmission mechanism" or the "Dirt protection mechanism". After selecting either existing or new Technology that can be adopted for the new product, we establish their priority using a mechanism-technology matrix. Now Parts can be deployed with the input from technology selected and product characteristics. The team can set the Costs of the parts. Parts for the product are either made or purchased from suppliers and the final product must be assembled: Process phases can be deployed and their priority calculated. At the end an analysis of Faults is implemented, in order to find out, from the customer's point of view, what the most serious faults that can occur are.

D. Maritan, *Practical Manual of Quality Function Deployment*,
DOI 10.1007/978-3-319-08521-0_3

3.1 Introduction

In the previous chapter we analysed the "Strategy" and "Customer" areas of the framework. We can summarise what we learned as follows: through a process of matrices and two field studies, *gemba* and a questionnaire survey, we passed from the definition of the adopted strategies (first three matrices, targets, core competencies, customer segments) to the deployment of the customer's requests (Demanded Qualities), and finally to creating a priority index using the Preplan matrix 4. These results are already very interesting, as they give a simple, clear order to product strategies, competitors benchmarking and customer analysis.

In this chapter we will focus on the technical area.

From an organisational point of view, the QFD team is made up of the same people as before, but now, R&D engineers and technical experts from the prototype and production departments have a more relevant role in giving their solutions and opinions.

3.2 What You Will Learn in This Chapter

In this chapter the rest of the framework phases ("c" to "g") will be discussed (Fig. 3.1):

- Product characteristics;
- Functions and Mechanisms;
- Innovation;
- Parts, Costs and Production process;
- Reliability.

The matrices used in these stages have the same mathematical approach as the strategic matrices 1, 2 and 3, seen in Chap. 2. Therefore you can concentrate on the QFD process flow and the unusual ways you connect the various product development stages, from design to selection of parts, costs, production process planning and reliability analysis. You can continue developing your QFD project, using the exercises as a guide to your work.

Remember: keep it simple (small matrices), your main objective is to test the framework through to the end, if possible.

In Fig. 3.2 the workflow of this chapter is presented from the information flow point of view. As you already know, priority is expressed a numerical value we obtain as an output of a matrix. We can also call the priority "weight" or "importance".

Two kinds of matrices will be used:

- Analytic Hierarchy Process (AHP) matrix;
- QFD correlation matrix.

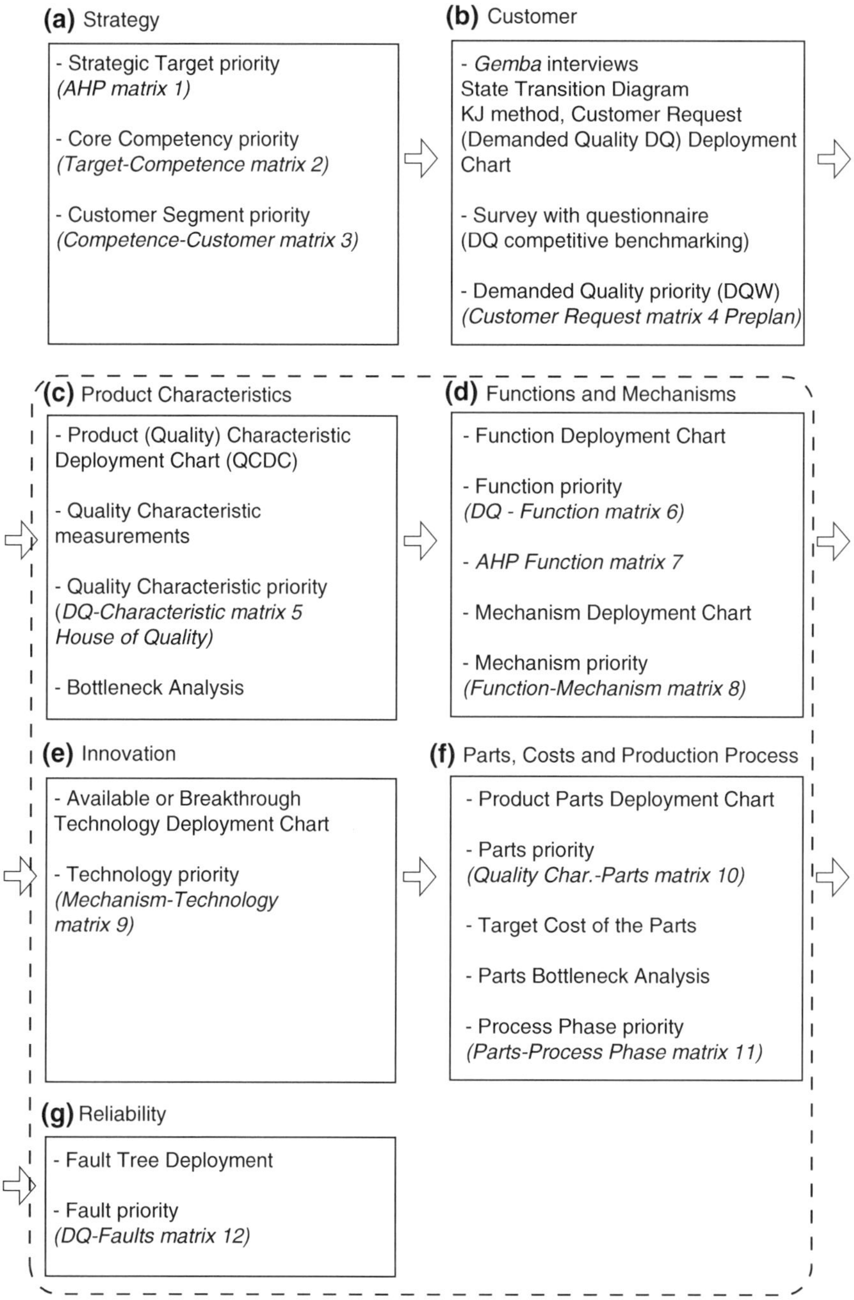

Fig. 3.1 Framework phases **c**–**g**, Product characteristics (phase **c**), Functions and Mechanisms (phase **d**), Innovation (Technologies) (phase **e**), Parts, Costs and Production process (phase **f**), Reliability (Faults) (phase **g**)

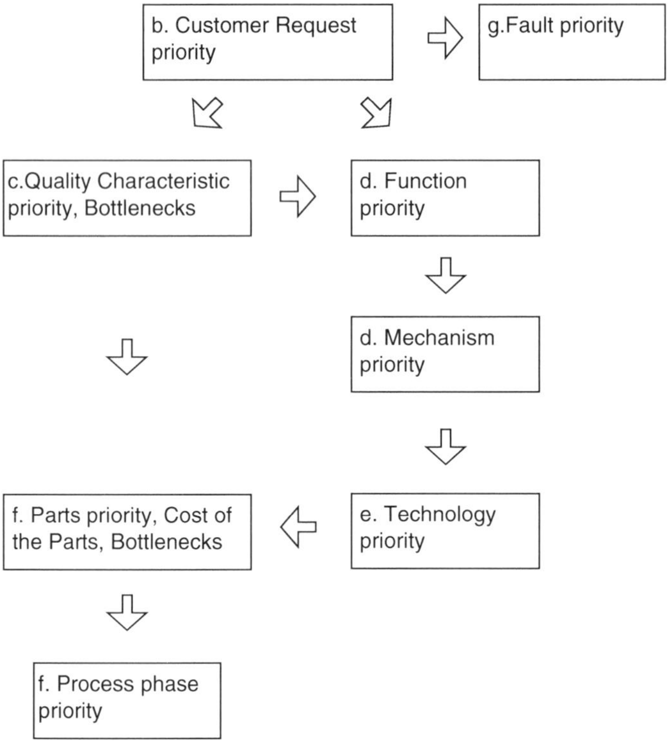

Fig. 3.2 Workflow of this chapter, from Customer requests to quality Characteristics, Functions, Mechanisms, Technologies, Parts, Costs, Process planning and Faults

The QFD team will have several brainstorming sessions making use of the Jiro Kawakita (KJ) method, to display, organise and select ideas and data.

The output from matrix 4 (Preplan, phase "b") is the customer requests priority, also called Demanded Quality Weight. This index is used in creating the House of Quality (phase "c") in order to obtain the priority of the product features, also called Quality Characteristics. Characteristics help to build the product Functions (an action fulfilling a product/service performance, i.e. for a bicycle, "It lights up the road ahead" or "It brakes"). Functions are weighted from both the customer and the technician's point of view (phase "d"). Functions then lead the team to the Mechanisms, macro-systems which carry out the product functions, for example, the "Steering mechanism" or the "Transmission mechanism" or the "Dirt protection mechanism". After selecting either existing or new Technology that can be adopted for the new product, we establish their priority using a mechanism-technology matrix (phase "e"). Now Parts can be deployed with the input from technology selected and product characteristics (phase "f"). The team can set the Costs of the parts. Parts for the product are either made or purchased from suppliers and the final product must be assembled: Process phases can be deployed and their priority calculated (phase "g").

At the end a parallel analysis of Faults is implemented, in order to find out, from the customer's point of view, what the most serious faults that can occur are.

Let's start with the case study.

3.3 Quality Characteristics and the House of Quality

An American Supplier Institute (ASI) definition presents Quality Function Deployment as a system to convert customer requests into company specifications during all product development process phases, such as marketing research, design, engineering, production, sales and maintenance (ASI 1987).

Effectively, in area "c" of our framework (Fig. 3.3), we will see that QFD allows Demanded Quality (DQ), i.e. the already ranked customer requests to be connected, to the product or service performance, also called "*Quality Characteristics*" (QC, Akao 1990) or "*Product Characteristics*".

In the next paragraphs we will learn how to deploy characteristics and link them to the DQ using a Request-Characteristic matrix 5: the so called "*House of Quality*" (HoQ). As we already know, House of Quality represents just one of the possible matrices used in this method. Certainly it is the most well-known, mainly for its ability to link two different contexts, the sales and the technical office, often considered as separate and conflicting, as they use different language styles and are made up of people with different cultural and educational backgrounds. Using the input from previous matrices we can obtain a weight for the technical characteristics of the products. We can, in other words, lead the design engineer during his choice of the performance target of the product, helping him during the unavoidable compromises he will have to accept (bottleneck).

3.3.1 Characteristics Deployment

What are Product Characteristics in QFD?

First of all they are not its parts or components; they describe its physical features and the results of tests and trials; they are the performance of a product or of a service. For instance they are its weight, dimensions, its speed, or the results of tests.

Characteristics must be *measurable* or at least *defined by a list* of possible alternatives.

In the next paragraphs the words "characteristic" and "feature" have the same meaning.

Starting from customer needs, the QFD inter-functional team has to identify items to design with features, which can be measured and checked, and respond to customer requests. This is a demanding phase, because it requires the translation of psychological expectations into technical objects.

Inside the working team technical people come to the fore, as they can benefit from the co-operation of the entire technical department and of the R&D department.

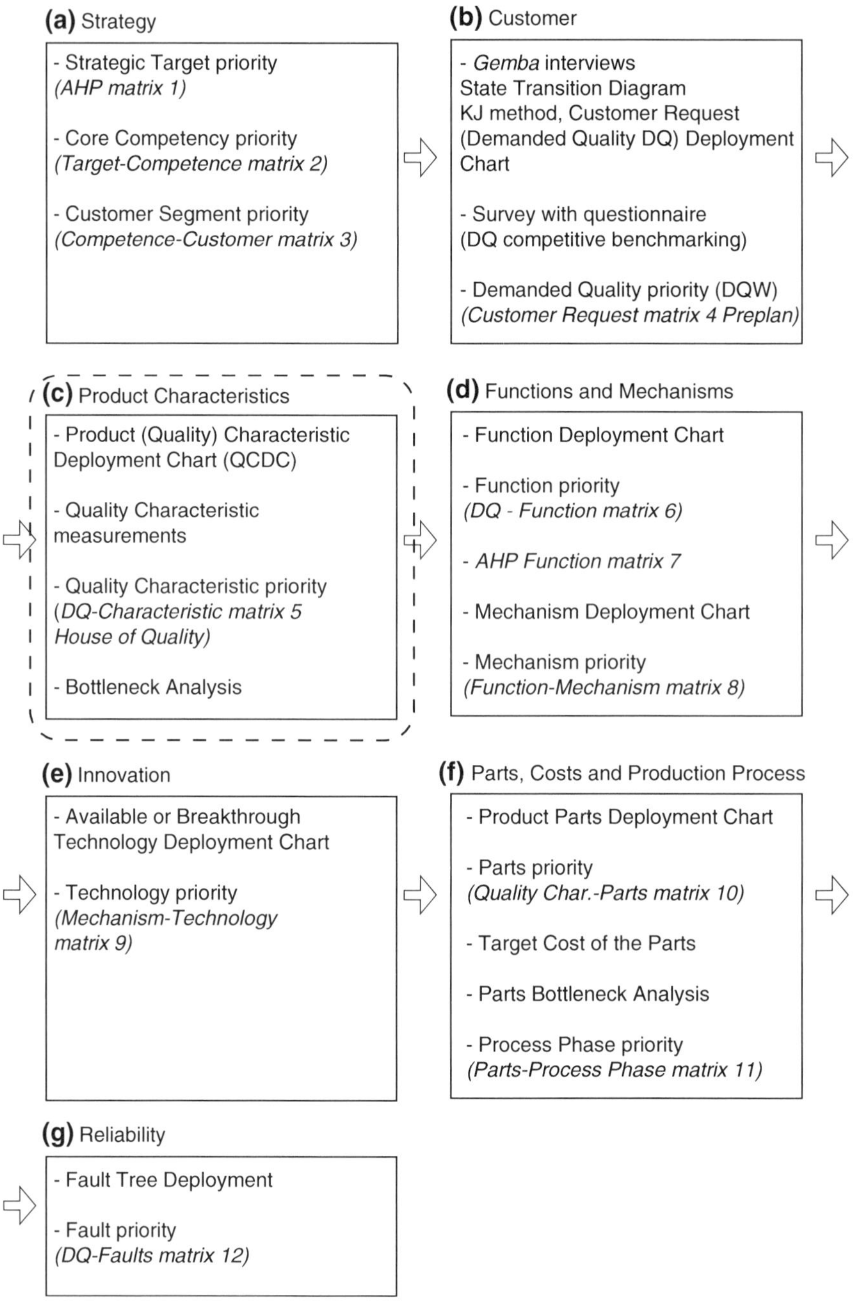
(a) Strategy
- Strategic Target priority
(AHP matrix 1)
- Core Competency priority
(Target-Competence matrix 2)
- Customer Segment priority
(Competence-Customer matrix 3)
(b) Customer
- Gemba interviews
State Transition Diagram
KJ method, Customer Request
(Demanded Quality DQ) Deployment
Chart
- Survey with questionnaire
(DQ competitive benchmarking)
- Demanded Quality priority (DQW)
(Customer Request matrix 4 Preplan)
(c) Product Characteristics
- Product (Quality) Characteristic
Deployment Chart (QCDC)
- Quality Characteristic
measurements
- Quality Characteristic priority
(DQ-Characteristic matrix 5
House of Quality)
- Bottleneck Analysis
(d) Functions and Mechanisms
- Function Deployment Chart
- Function priority
(DQ - Function matrix 6)
- AHP Function matrix 7
- Mechanism Deployment Chart
- Mechanism priority
(Function-Mechanism matrix 8)
(e) Innovation
- Available or Breakthrough
Technology Deployment Chart
- Technology priority
(Mechanism-Technology
matrix 9)
(f) Parts, Costs and Production Process
- Product Parts Deployment Chart
- Parts priority
(Quality Char.-Parts matrix 10)
- Target Cost of the Parts
- Parts Bottleneck Analysis
- Process Phase priority
(Parts-Process Phase matrix 11)
(g) Reliability
- Fault Tree Deployment
- Fault priority
(DQ-Faults matrix 12)

Fig. 3.3 Framework phase **c**, Product Characteristics

The powerful tool of the KJ method (see Chap. 2) is used again. KJ makes this technical experience clearer, by involving the company and the team in selection of characteristics.

In this case the starting point is not the customer's raw data but the Demanded Qualities. Sentences are re-worked, changed into technical friendly language, and then compared with the other sentences, becoming measurable Product/Service Characteristics; afterwards they are listed in a hierarchical structure with few levels (maximum 3), the Quality Characteristic Deployment Chart (QCDC).

Royal bicycle QCDC is represented with the following characteristics (one level of detail):

- Braking noise level;
- Chain noise level;
- Hub noise level;
- Bicycle weight;
- Available colours;
- Accessories;
- Component labelling;
- Bike setting;
- Paintwork durability test;
- Warranty;
- Product delivery time;
- Lighting efficiency;
- Braking distance.

Remember that these characteristics are just *one of the possible unlimited results of KJ sessions.*

Teamwork helps maintain a strong sense of originality (see Chap. 1).

In this specific case, we notice that some characteristics are clearly and easily identified, others are more generic and difficult to measure. A complete description of the measurement test, which will unambiguously determine the value of each characteristic, is needed for all of them. Description of each single test can be more or less detailed, according to the level of know-how of the group, or to the amount of time/resources you want to devote to this process. As an alternative to the numerical value you can measure a characteristic by making a choice between certain items (for instance: Presence of User Manual: YES, NO).

As an example, we analyse the description of the measurement test of "Braking noise".

The test can be an evaluation of the bicycle noise level during braking, both on wet or dry surfaces. The test is performed on a dynamic work bench and a sound-meter with specific features is used. A few sprayers on the bench wet the wheel during rotation, at a distance of 150–200 mm, using de-mineralised water at room temperature. The test measures the SPL (Sound Pressure Level) value generated by the bicycle in Decibels (dB). The final value is a weighted average of the two tests:

$$dB = 1/4 \text{ dBWet } + 3/4 \text{ dBDry}$$

The lower the SPL is, the better the softness of the braking system. The weights refer to the frequency of use under wet or dry conditions.

Exercise It is time to start a new KJ session for product/service characteristics. Start from the list of customer requests and create a list of no more that 10 characteristics using the KJ method (see Chap. 2). Remember that they have to be measurable or, at least, selectable from a list of options. Take time to discuss the procedure of testing each characteristic and measure each of them. Is the KJ session different from the previous one, focused on the customer? Why? How much time did you spend on this phase? Was it faster than KJ for customers (learning curve) or not? Why?

3.3.2 Request—Characteristic Matrix 5 (House of Quality—HoQ)

The House of Quality has a very similar layout to the strategic matrices 2 and 3, referred to as Core competency and Customers (see Chap. 2). If we have a look at the Royal Project matrix 5 (Fig. 3.4), on the left hand side we can see that the

CUSTOMER REQUEST – CHARACTERISTIC MATRIX 5 (HOUSE OF QUALITY)															
		CHARACTERISTIC DEPLOYMENT													
		Comfort				Customisation			Service and after sales				Safety		
CUSTOMER REQUEST (DEMANDED QUALITY) DEPLOYMENT	Customer request priority (DQW)	Braking noise level	Chain noise level	Hub noise level	Bicycle weight	Available colours	Accessories	Component labelling	Bike setting	Paintwork durability test	Warranty	Product delivery time	Lighting efficiency	Braking distance	
Feeling of refinement and classic style	11.16%	∴	⊘	⊘		∴	⊙	⊙			⊘		∴		
Saving time in assembling	8.82%				⊘				⊙		∴	⊘			
Riding a status symbol	9.92%	⊘	⊘	⊘	∴	∴	⊙	⊙	∴	⊘	⊙		⊘	⊘	
Make an investment for the future	12.54%						⊘	∴		⊙	⊙				
Absence of back stress	5.16%				⊘		∴							∴	
Regular progression in braking	12.27%	⊙	∴	∴	∴				∴					⊙	
Feeling of silence while riding	7.95%	⊘	⊙	⊙										∴	
Feeling of fluidity while pedalling	6.78%		⊙	⊘	⊘				⊘						
Need for sports style	4.82%	⊘			⊙		⊘	∴					⊘	⊘	
I feel different from the other cyclists	11.87%	∴	∴	∴		⊙	⊙	⊙		∴	⊘				
Feeling of craftsmanship	8.68%						∴	⊙	∴		∴				
Technical benchmarking	Unit of measurement	dB	dB	dB	kg	no. colours	no. access.	no. compon.	min	hours	years	days	metres	metres	
Technical benchmarking	Royal	50	37	46	17	1	7	5	45	48	2	10	12	4	
Technical benchmarking	Competitor A	40	39	40	15	3	10	3	25	48	3	15	13	3	
Technical benchmarking	Competitor B	36	43	40	16.5	2	15	5	25	48	3	15	12	4	
Technical benchmarking	Technical Targets	36	35	38	15	3	20	7	25	72	5	7	15	3	
	Difficulty (1=very easy, 5=very difficult)	4	2	1	3	2	2	3.5	4	5	4	3	2	2	Total
	Absolute weight	2.02	2.20	1.79	1.28	1.28	3.63	3.92	1.31	1.54	2.89	0.26	0.55	1.68	24.35
	Characteristic Priority (QCW)	*8.28%*	*9.04%*	*7.37%*	*5.25%*	*5.25%*	*14.89%*	*16.10%*	*5.36%*	*6.34%*	*11.86%*	*1.09%*	*2.28%*	*6.89%*	100.00%

Fig. 3.4 Royal bicycle House of Quality matrix 5

heading each row comes from the Demanded Quality Deployment Chart (DQDC), and links each DQ with its relative weight (DQW—Demanded Quality Weight, second column). Column headings come from the Quality Characteristic Deployment Chart (QCDC, two levels). The matrix is filled in by answering the question about when and how much each Characteristic (remember that they are also called "Hows") satisfies the different Customer Requests ("Whats").

Symbols of correlation (same as for the strategic matrices) can be inserted into the cells or these can remain empty if there is no link.

The symbols traditionally used are:

⊙ = strong correlation
⊘ or ○ = medium correlation
□ or Δ = weak correlation

Weights corresponding to symbols are "9" for strong correlation, "3" for medium correlation and "1" for weak correlation; an empty cell corresponds to "0" value.

We know that the absolute weight of the characteristics is calculated with the Independent Scoring Method (see Chap. 2).

The weighted sums in the columns are made relative, obtaining the "Quality *Characteristic priority*", also called "Quality Characteristic Weight" (QCW). The priority can now be sorted and shown in a descending order diagram (Fig. 3.5). This index is a very important result for the design engineer, because it gives direction to the project, by assigning priority to the characteristics with a higher score.

There are further interesting evaluations in the "foundation" of House of Quality: the *technical benchmarking* and the *Bottleneck Analysis*.

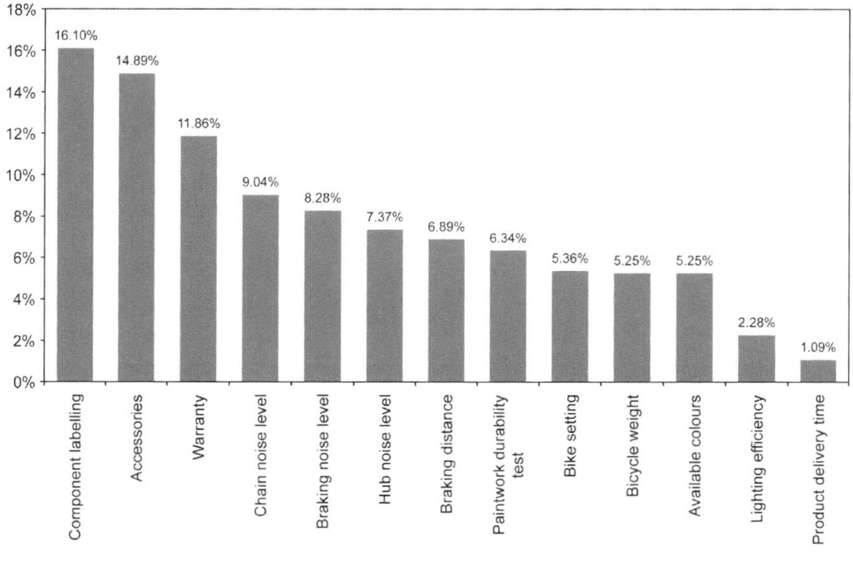

Fig. 3.5 Royal Quality Characteristic priority

Column by column, technical benchmarking shows the results of the tests for all the products we want to consider, usually our product and that of one or two competitors. The QFD team then fixes future targets for each characteristic, with the aim of improving the best performance obtained from the tests. For example, the current warranty on Royal is for 2 years, the competitors' warranty is for 3 years. Thus the target warranty time for the new product is 5 years.

Akao called these targets "Quality of Design" (Akao 1990). However this term can create some confusion, so I suggest calling them "*Technical Targets*".

To summarise, for each characteristic the output is:

- a relative weight, i.e. characteristics priority;
- a result set by tests;
- a technical target.

Exercise Let's create the House of Quality for your QFD project. Take the Royal case study as a guideline. Where possible, set up the formulas needed to automate computation in the software spreadsheet. Follow these stages:

(a) create a new matrix; start from matrix 2 or 3, because the settings are identical, what changes is the number of rows and columns and the deployment charts (requests and characteristics);
(b) copy the request deployment chart and the priority from matrix 4 (Preplan);
(c) pay attention to the absolute weight programming (here the risk of making mistakes is higher);
(d) create the technical benchmarking area (2–3 rows, depending on the number of competitors) and determine a first plausible technical target;
(e) take your time to check, test and measure the characteristics;
(f) characteristics priority should be calculated automatically;
(g) create the histogram and comment on it.

Take note of the time spent and energy used during this phase.

Now the working team estimates the "*Difficulty*" (D) in reaching the technical target of the characteristic.

So another row is then added, where evaluation of the difficulty is assessed by using a five point Likert scale.

1 = very easy
2 = quite easy
3 = a bit difficult
4 = rather difficult
5 = very difficult

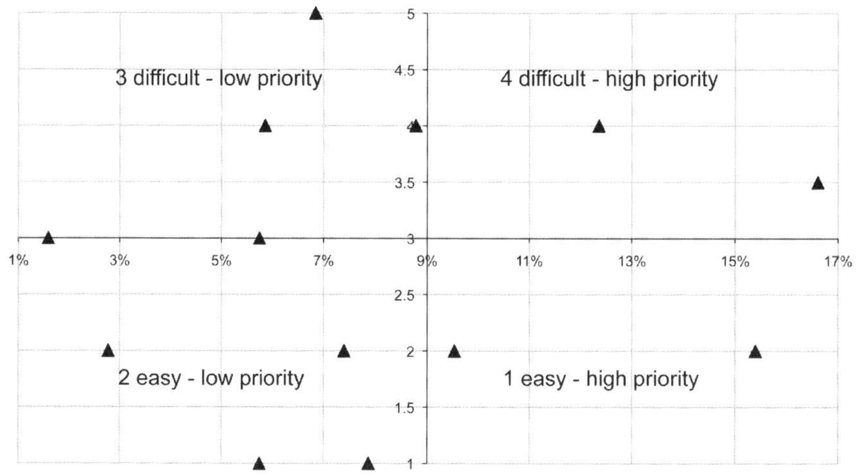

Fig. 3.6 Product Characteristic bottleneck analysis, Royal bicycle

I also suggest inserting intermediate values between two assessments, with a possible evaluation of half a point, for instance 3.5 means that reaching the technical target is between "a bit difficult" and "rather difficult".

The evaluation is made by the team who give a specific meaning to the word "difficult", by taking into consideration both the management and technical aspects (if necessary you can split them and create two different assessments). The final decision is not taken with an average value, but the team must find a compromise between the conflicting opinions of the members, therefore brainstorming becomes more and more demanding.

Estimated Difficulty D in reaching a specific target and Characteristic Weight QCW are two wonderful parameters very useful for *bi-dimensional mapping of the characteristics* (Bottleneck Analysis).

In the X axis of the diagram we indicate the Characteristic Weight, in the Y axis the Difficulty level (Fig. 3.6). Each point is represented by the pair of numbers (QCW, D).

We can now divide the diagram into four sections.

1. First section, bottom right hand side, includes characteristics that are important and quite easy to improve with reference to the Technical Target; we can name this "Easy—High priority". In the Royal bicycle case, the characteristics: "Accessories", with a very high score in the relative weight, and "Chain sound level" belong in this section The R&D technicians will first concentrate on this improvement, obtaining the maximum result with a minimum of effort.
2. The other section, bottom left hand side, includes characteristics with very low importance but easy to implement. The team can work on this section too, which we can call "Easy—Low priority", being aware that the priorities of these product/service performances are very low.

3. The third section, the top left one, includes characteristics with low priority and very challenging technical targets: these issues, "Difficult—Low priority", need to be revised, reducing expectations of improvement compared to competitors, moving the characteristics from the third to the second section (see arrow in the diagram); by relocating targets we make them easier to reach. In this case the technical approach has to be partially conservative, sometimes maintaining the status quo of the product.
4. The fourth section is dedicated to Bottlenecks. They represent very important features in order to achieve full satisfaction of the Customer Requests, but with/ have challenging technical Targets. I call this area "Difficult—High priority". "Component Labelling" and "Warranty" are therefore bottlenecks. These characteristics require in-depth analysis of tests and targets, with the aim of finding a possible solution to the impasse generated by this section.

Exercise Now add a row at the bottom of your House of Quality. Fill the cells with a 1–5 scale of the difficulty estimated to reach the Technical Targets. Then, with the help of your spreadsheet software, create a graph like Fig. 3.6. Study the deployment of the points. What are the bottlenecks in your project? What are your "easy–important" characteristics?

3.4 Functions and Product Mechanisms

Most of the QFD projects stop with the House of Quality, but, as we'll see, it can be very useful to proceed and create more matrices and go on as far as the production process.

We can find out, as suggested by Akao and Matzumoto (Akao 1990),

- which product functions express the identified characteristics;
- which mechanisms are needed to carry out product functions.

See Fig. 3.7.

3.4.1 Function Deployment

A function is an action fulfilling a product/service performance, normally described by a subject and a verb, a measurable noun. It mustn't be confused with the word *kino* contained in the method name Quality Function Deployment, which refers to the functions performed by the company organisation to deploy quality throughout the product development process.

A Function is therefore what a product makes, not the benefit obtained from it. Mazur (1999), for instance, suggests avoiding the expression "it allows for...".

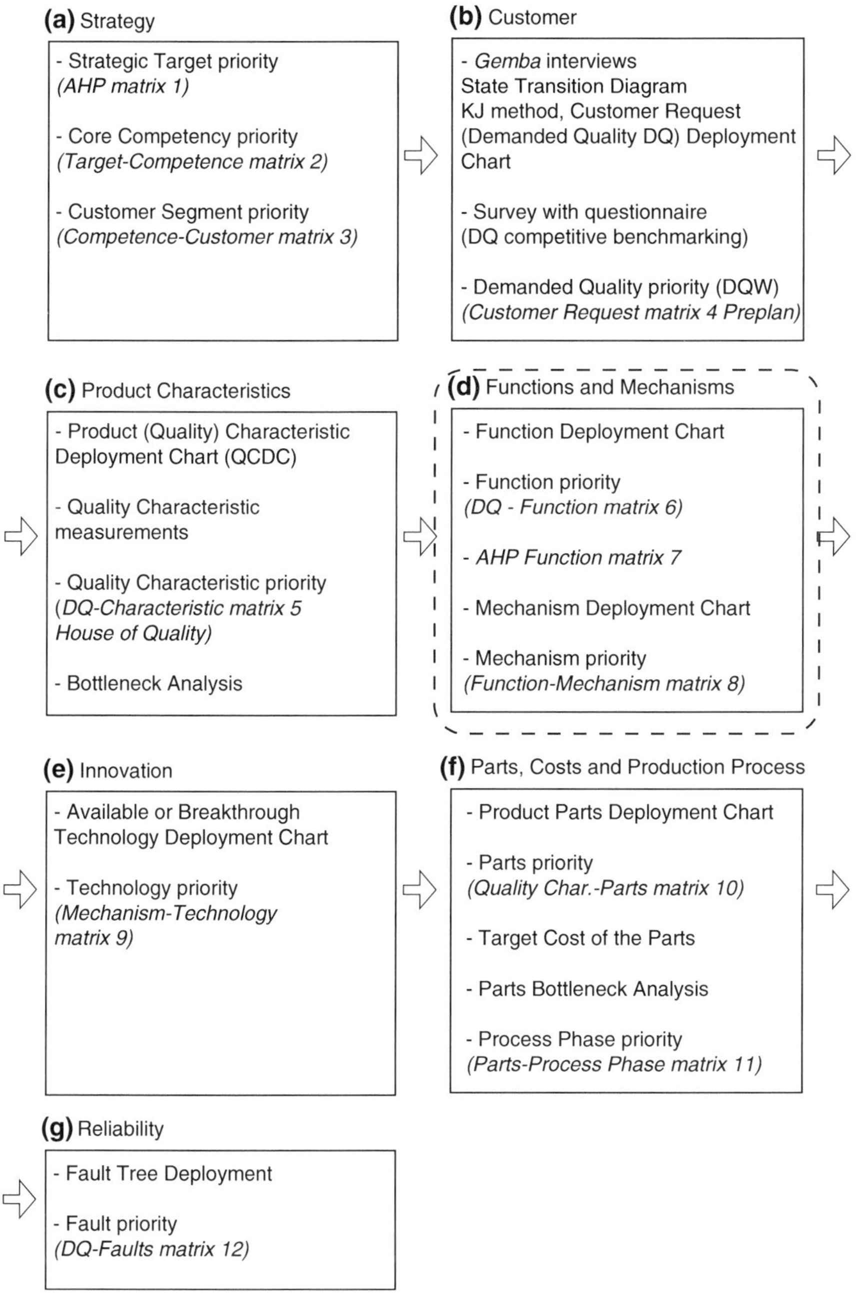
(a) Strategy
- Strategic Target priority
(AHP matrix 1)
- Core Competency priority
(Target-Competence matrix 2)
- Customer Segment priority
(Competence-Customer matrix 3)
(b) Customer
- Gemba interviews
State Transition Diagram
KJ method, Customer Request
(Demanded Quality DQ) Deployment
Chart
- Survey with questionnaire
(DQ competitive benchmarking)
- Demanded Quality priority (DQW)
(Customer Request matrix 4 Preplan)
(c) Product Characteristics
- Product (Quality) Characteristic
Deployment Chart (QCDC)
- Quality Characteristic
measurements
- Quality Characteristic priority
(DQ-Characteristic matrix 5
House of Quality)
- Bottleneck Analysis
(d) Functions and Mechanisms
- Function Deployment Chart
- Function priority
(DQ - Function matrix 6)
- AHP Function matrix 7
- Mechanism Deployment Chart
- Mechanism priority
(Function-Mechanism matrix 8)
(e) Innovation
- Available or Breakthrough
Technology Deployment Chart
- Technology priority
(Mechanism-Technology
matrix 9)
(f) Parts, Costs and Production Process
- Product Parts Deployment Chart
- Parts priority
(Quality Char.-Parts matrix 10)
- Target Cost of the Parts
- Parts Bottleneck Analysis
- Process Phase priority
(Parts-Process Phase matrix 11)
(g) Reliability
- Fault Tree Deployment
- Fault priority
(DQ-Faults matrix 12)

Fig. 3.7 Framework phase **d**, Functions and Mechanisms

Table 3.1 Function Deployment Chart, Royal bicycle; in *italics* you see the Functions critical path with the essential function "It moves rear wheel"

It moves people and object	*It moves people*	*It runs on the road*	It steers front wheel
			It moves rear wheel
		It lights the road	It lights the road
		It protects rider	It rings the bell
			It brakes
		It gives comfort	It serves as seating for cyclist
			It keeps cyclist clean
	It carries objects	It carries objects	It carries objects

Once again, the KJ meetings help to determine the so-called *Function Deployment Chart* (FDC), i.e. the Deployment of the Functions carried out by the product according to an affinity tree divided into several levels. Table 3.1 shows the FDC (four levels) for the Royal bicycle; note the italic style path, which we can call "function critical path": the essential function for the product is identified here, in this case "it moves rear wheel", which allows the bicycle to cover a distance and in more general terms, allows people or objects to be moved.

This is a very important phase in the QFD process, because with these product functions we can fill the empty spaces left by the Customer in his general evaluation of the product.

As you read in Chap. 2, in the Kano model, which describes the typologies of qualities in a product, there are some qualities that inevitably have to be present in a product. If not, potentially their absence can lead to disappointment on the part of the customer. He probably won't buy the product or the service. These are the so-called "Must-be" qualities (Kano et al. 1984). On the other hand, if the must-be qualities are present, the client may be indifferent. In the automotive industry "must-be" qualities can be the presence of wheels or nowadays ABS or airbag.

"Must be" qualities are not expressed in *gemba* interviews. Functions analysis allows us to focus on these qualities too. In other words, technicians must ask themselves: what are the basic functions of the product or service?

The FDC is now linked to the Customer Requests or to the product Characteristics. Probably if we consider the end user's position (Business to Customer BtoC), a link with the requests is much more interesting, although the Characteristics-Functions matrix calculates the weight of the product functions from a technical point of view. The two matrices should indicate similar results, at least for the items with higher scores; if there are big discrepancies it is better to verify the calculation procedure (ISM), together with a second check of the Function list with all its symbols.

As an example we create the *Customer Request—Function* matrix 6 for the Royal bicycle (Fig. 3.8).

CUSTOMER REQUEST – FUNCTION MATRIX 6										
	Customer request priority (DQW)	FUNCTION DEPLOYMENT								
		It moves people and objects								
		It moves people							It carries objects	
		It runs on the road		It lights the road	It protects rider		It gives comfort		It carries objects	
CUSTOMER REQUEST (DEMANDED QUALITY) DEPLOYMENT		It steers front wheel	It moves rear wheel	It lights the road	It rings the bell	It brakes	It serves as seating for cyclist	It keeps cyclist clean	It carries objects	
Feeling of refinement and classic style	11.16%				⁂	⊘		⊘	⊘	
Saving time in assembling	8.82%	⁂	⁂			⊘				
Riding a status symbol	9.92%			⊘	⁂	⁂	⊘	⊘	⁂	
Make an investment for the future	12.54%								⁂	
Absence of back stress	5.16%		⁂				⊙			
Regular progression in braking	12.27%					⊙				
Feeling of silence while riding	7.95%	⁂	⊙			⊙	⁂			
Feeling of fluidity while pedalling	6.78%		⊙				⁂		⁂	
Need for sports style	4.82%		⁂			⁂				
I feel different from the other cyclists	11.87%				⊘	⊙		⊘	⁂	
Feeling of craftsmanship	8.68%									Total
	Absolute weight	0.17	1.51	0.30	0.57	3.64	0.91	0.99	0.75	8.83
	Function Priority	*1.90%*	*17.16%*	*3.37%*	*6.42%*	*41.19%*	*10.31%*	*11.20%*	*8.45%*	100.00%

Fig. 3.8 Royal Function-Customer Request matrix 6

Functions that will be used in the matrix are:

- It steers front wheel;
- It moves rear wheel;
- It lights the road;
- It rings the bell;
- It brakes;
- It serves as seating for cyclist;
- It keeps cyclist clean;
- It carries objects.

Matrix 6 is completed evaluating the ability (strong, medium, weak or no correlation) of the functions to fulfil the customer request.

We can see, for instance, that "It moves rear wheel" is strictly correlated to "Feeling of silence while riding" and to "Feeling of fluidity while pedalling". Some links can depend on the presence of a previous version of the product; this is the explanation of the strong link between "It brakes" and the DQ "I feel different from the other cyclists": this product, in fact, is characterised by an unusual "integral braking system" of the brake levers, which activates the braking pliers on both wheels.

As for previous matrices, it is possible to draw a function priority graph, obtaining a histogram like Fig. 3.9. We can deduct that function "It brakes" is

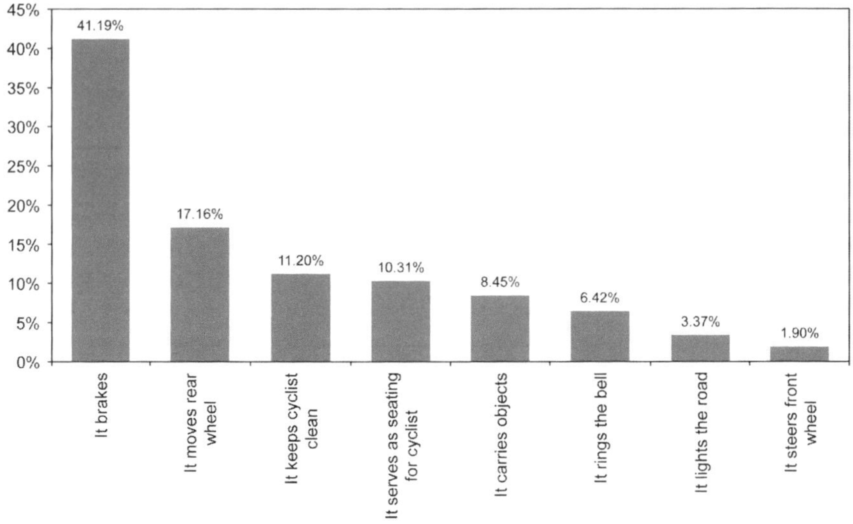

Fig. 3.9 Royal Functions priority

considered crucial for the overall satisfaction of the customer, followed by the function of moving the rear wheel (critical path) and keeping the rider clean.

Up to now we have analysed the Functions considering their impact on the Customer Requests.

It is also useful to analyse them from an engineering point of view. We apply the AHP methodology (matrix 7 of our framework), correlating the Functions among themselves with the aim of identifying the most important ones (Akao 1990). See Figs. 3.10 and 3.11.

We immediately notice that in this case AHP confirms the results obtained from the link with the Customer Requests as far as concerns the first two matrices, "It moves rear wheel" and "It brakes", while the function "It steers front wheel", ranked third from a technical point of view, has a very low score according to customer opinion: the customer seems to take it for granted for such a product. The function "It keeps cyclist clean" has an opposite ranking, compared to the previous function, so it has a very low score in AHP and a very high score in Requests-Functions matrix.

Exercise Let's create your Function deployment chart and Function matrices. Take the Royal case study as a guideline. Where possible, set up the formulas needed to automate computation in the software spreadsheet. Follow these stages.

(a) Create the Function Deployment Chart (few basic functions); do not hesitate to ask friends, not involved in your team, to specify what the functions of your product or service are. Remember that you may have to identify "must-be" functions.

(b) Create Function-Customer Request matrix 6.
(c) Create Function AHP matrix 7.
(d) Function priorities should be calculated automatically.
(e) Create the histograms and comment on them.

Are you lost in the forest of matrices? Go to framework Figs. 3.1 and 3.2. Take note of the time spent and energy used during this phase.

AHP FUNCTION MATRIX N. 7

1 = the row is equally important as the column
3 = slightly more important
5 = more important
7 = much more important
9 = very much more important
0,33 = slightly less important
0,20 = less important
0,14 = much less important
0,11 = very much less important

FUNCTION DEPLOYMENT		FUNCTIONS: 1 It steers front wheel	2 It moves rear wheel	3 It lights the road	4 It rings the bell	5 It brakes	6 It serves as seating for cyclist	7 It keeps cyclist clean	8 It carries objects	Normalised assessment: *1*	*2*	*3*	*4*	*5*	*6*	*7*	*8*	Absolute weight	FUNCTION AHP PRIORITY (%)
1	It steers front wheel	1.00	0.20	5.00	7.00	1.00	5.00	7.00	5.00	*0.13*	*0.07*	*0.20*	*0.21*	*0.26*	*0.30*	*0.21*	*0.15*	1.52	**19.04%**
2	It moves rear wheel	5.00	1.00	7.00	9.00	1.00	5.00	7.00	7.00	*0.63*	*0.34*	*0.28*	*0.26*	*0.26*	*0.30*	*0.21*	*0.21*	2.50	**31.30%**
3	It lights the road	0.20	0.14	1.00	3.00	0.14	0.33	3.00	1.00	*0.03*	*0.05*	*0.04*	*0.09*	*0.04*	*0.02*	*0.09*	*0.03*	0.38	**4.73%**
4	It rings the bell	0.14	0.11	0.33	1.00	0.20	0.20	1.00	0.33	*0.02*	*0.04*	*0.01*	*0.03*	*0.05*	*0.01*	*0.03*	*0.01*	0.20	**2.53%**
5	It brakes	1.00	1.00	7.00	5.00	1.00	5.00	7.00	7.00	*0.13*	*0.34*	*0.28*	*0.15*	*0.26*	*0.30*	*0.21*	*0.21*	1.88	**23.49%**
6	It serves as seating for cyclist	0.20	0.20	3.00	5.00	0.20	1.00	7.00	9.00	*0.03*	*0.07*	*0.12*	*0.15*	*0.05*	*0.06*	*0.21*	*0.27*	0.95	**11.93%**
7	It keeps cyclist clean	0.14	0.14	0.33	1.00	0.14	0.14	1.00	3.00	*0.02*	*0.05*	*0.01*	*0.03*	*0.04*	*0.01*	*0.03*	*0.09*	0.27	**3.42%**
8	It carries objects	0.20	0.14	1.00	3.00	0.14	0.11	0.33	1.00	*0.03*	*0.05*	*0.04*	*0.09*	*0.04*	*0.01*	*0.01*	*0.03*	0.29	**3.56%**
	Total	7.88	2.93	24.66	34.00	3.82	16.78	33.33	33.33	*1.00*	*1.00*	*1.00*	*1.00*	*1.00*	*1.00*	*1.00*	*1.00*	8.00	100.00%

Fig. 3.10 Royal Function AHP matrix 7

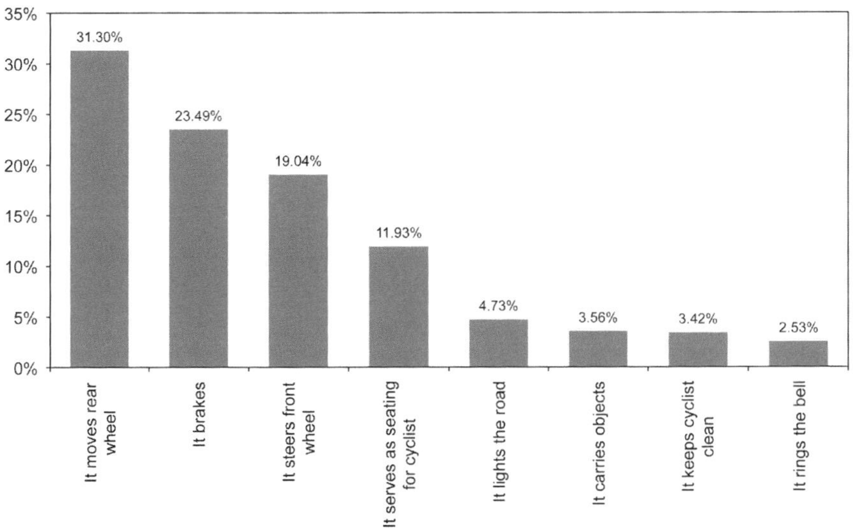

Fig. 3.11 Function technical priority, Royal bicycle

3.4.2 Mechanisms Deployment

Mechanisms represent the macro-systems that carry out the product functions.

We need to clarify that mechanisms are not the parts of the product. In fact we are developing a new product and, during this phase, the choices regarding innovation of mechanisms have not been made yet. Later, depending on these choices, the parts of the system will be completely defined and they will be ready to be weighted in a QFD matrix.

The development of mechanisms and their priority help decide where to work harder on innovation, which will lead, from mechanism to mechanism, to finding new technological solutions and new parts, maybe not existing in the previous product or in those of the competitors.

In this phase the QFD team tries to identify the mechanisms. New KJ sessions lead to the definition of the Mechanism Deployment Chart (MDC). Royal bicycle MDC shows the following mechanisms (one level of detail):

- Steering mechanism;
- Transmission mechanism;
- Lighting mechanism;
- Saddle and frame system;
- Braking system;
- Dirt protection mechanism;
- Carriage system.

Normally the work is not very detailed, as we are now defining macro-systems, which carry out the previously identified functions; we are not interested in being detailed; the parts deployment can be detailed in a later phase.

The MDC is linked to the functions through the Function-Mechanism matrix 8, see Fig. 3.12.

If we look at the links inside the matrix, for instance, the "Transmission system" is strictly and strongly related to the functions "It moves rear wheel" and "It keeps cyclist clean". The same, the "Saddle and frame system" is strictly linked to "It serves as seating for cyclist" and has a weak or medium link with other functions.

Mechanism priority can be ranked in a diagram (Fig. 3.13). If we have a look at the results, we notice that the "Braking system" and the "Transmission mechanism" obtain the highest scores, followed by the chassis and protection from dirt.

Now we can take advantage of the Mechanism analysis, we have just completed, for innovation which will be a feature of our product.

Exercise Let's create your Mechanism Deployment Chart and Mechanism matrix 8. Take the Royal case study as a guideline. Where possible, set up the formulas needed to automate calculation in the software spreadsheet. Follow these stages.

FUNCTION - MECHANISM MATRIX 8									
		MECHANISM DEPLOYMENT							
FUNCTION DEPLOYMENT	Function priority	Steering mechanism	Transmission mechanism	Lighting mechanism	Saddle and frame system	Braking system	Dirt protection mechanism	Carriage system	
It steers front wheel	1.90%	⊙			⁂				
It moves rear wheel	17.16%		⊙		⁂	⁂	⊘		
It lights the road	3.37%	⊘		⊙	⁂				
It rings the bell	6.42%				⊘				
It brakes	41.19%		⁂		⊘	⊙			
It serves as seating for cyclist	10.31%		⊘		⊙		⁂		
It keeps cyclist clean	11.20%		⊙		⁂		⊙		
It carries objects	8.45%							⊙	Total
	Absolute weight	0.27	3.27	0.30	2.69	3.88	1.63	0.76	12.81
	Mechanism priority	***2.13%***	***25.56%***	***2.37%***	***21.02%***	***30.28%***	***12.70%***	***5.94%***	100.00%

Fig. 3.12 Royal Function-Mechanism matrix 8

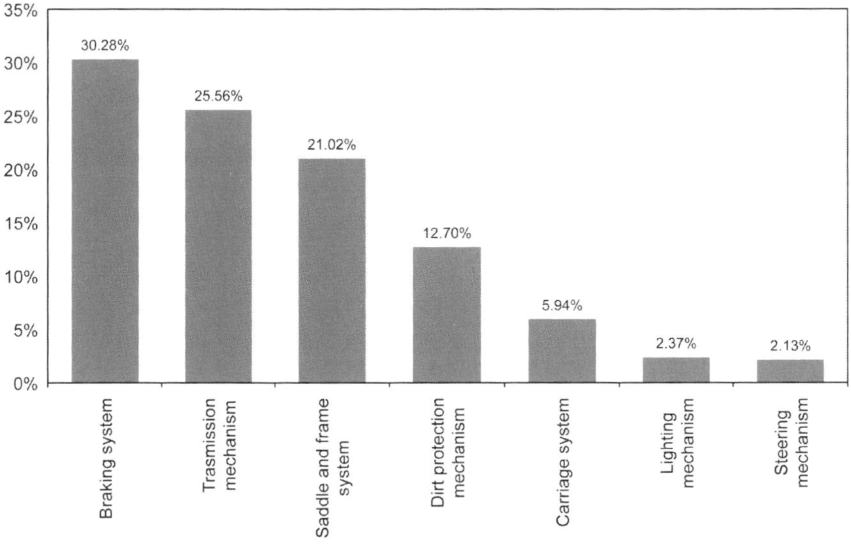

Fig. 3.13 Royal Mechanism priority

(a) Create the Mechanisms Deployment Chart; mechanisms are not parts of the product. Start from the Functions Deployment and use a quick KJ session. Find a maximum of 10 mechanisms. Do you feel comfortable using KJ?

(b) Create Functions—Mechanism matrix 8.
(c) Mechanism priorities should be calculated automatically.
(d) Create the histogram and comment on it.

Take note of the time spent and energy used during this phase.

3.5 Innovation

Technological innovation (phase "e" of the framework, see Fig. 3.14) can be introduced:

- considering available technology or the one currently under development;
- indicating radical solutions, breakthrough.

The Royal bicycle case study only uses technology of the first type. By using the KJ method together with a new QFD matrix, we can select possible innovations which characterise the new product/service. In some situations the technology available is not enough. Some mechanisms, that Akao (1990) calls bottleneck mechanisms, require radical innovation, such as a technological breakthrough.

With one session of the Jiro Kawakita method for the Royal bicycle project the working team indicates a group of technical solutions that the present product does not have, derived from the bike market or from other similar fields.

A *Technology Deployment Chart* is then created (one level).

- Led lights;
- Disc brakes;
- Light frame (aluminium, titanium?);
- Rear hub gear;
- New paintwork process;
- Front suspension fork;
- Bag hooks instead of rack.

To give an example we will comment on a few innovations that could be possible with the available technology. The led bicycle lighting can replace traditional lighting, from the light generation point of view, although the external look of the lighting system should remain classic and unchanged. The lightweight frame (aluminium or titanium for the top range version) will allow for substantial weight reduction compared to the original steel chassis, again the traditional look of the shape and painting should be maintained. Disc brakes and changing to a hub gear are probably the two main innovations and they represent the main risk on a product like this. While gear changing is carried out by a very discretely disguised

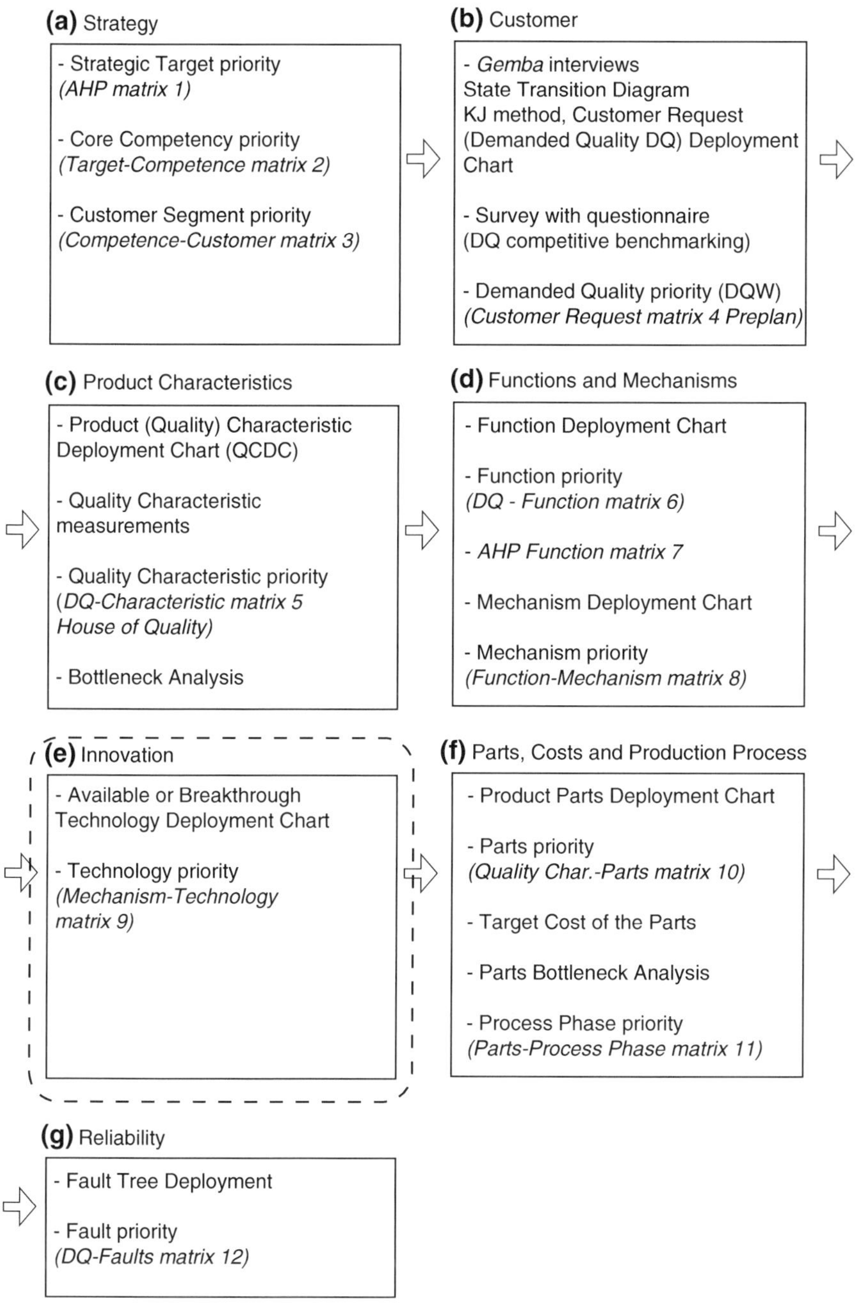
(a) Strategy
- Strategic Target priority
(AHP matrix 1)
- Core Competency priority
(Target-Competence matrix 2)
- Customer Segment priority
(Competence-Customer matrix 3)
(b) Customer
- Gemba interviews
State Transition Diagram
KJ method, Customer Request
(Demanded Quality DQ) Deployment
Chart
- Survey with questionnaire
(DQ competitive benchmarking)
- Demanded Quality priority (DQW)
(Customer Request matrix 4 Preplan)
(c) Product Characteristics
- Product (Quality) Characteristic
Deployment Chart (QCDC)
- Quality Characteristic
measurements
- Quality Characteristic priority
(DQ-Characteristic matrix 5
House of Quality)
- Bottleneck Analysis
(d) Functions and Mechanisms
- Function Deployment Chart
- Function priority
(DQ - Function matrix 6)
- AHP Function matrix 7
- Mechanism Deployment Chart
- Mechanism priority
(Function-Mechanism matrix 8)
(e) Innovation
- Available or Breakthrough
Technology Deployment Chart
- Technology priority
(Mechanism-Technology
matrix 9)
(f) Parts, Costs and Production Process
- Product Parts Deployment Chart
- Parts priority
(Quality Char.-Parts matrix 10)
- Target Cost of the Parts
- Parts Bottleneck Analysis
- Process Phase priority
(Parts-Process Phase matrix 11)
(g) Reliability
- Fault Tree Deployment
- Fault priority
(DQ-Faults matrix 12)

Fig. 3.14 Framework phase **e**, Innovation

MECHANISM – TECHNOLOGY MATRIX 9									
		TECHNOLOGY DEPLOYMENT							
MECHANISM	Mechanism priority	Led lights	Disc brakes	Light frame (aluminium, titanium?)	Rear hub gear	New paintwork process	Front suspension fork	Bag hooks instead of rack	
Steering mechanism	2.13%			⊘			⊙		
Transmission mechanism	25.56%		⁂		⊙		⁂		
Lighting mechanism	2.37%	⊙							
Saddle and frame system	21.02%			⊙		⊘	⁂	⁂	
Braking system	30.28%		⊙		⁂				
Dirt protection mechanism	12.70%				⊘	⁂		⊘	
Carriage system	5.94%			⊘				⊙	Total
	Absolute weight	0.21	2.98	2.13	2.98	0.76	0.66	1.13	10.85
	TecHnology priority	*1.97%*	*27.47%*	*19.66%*	*27.50%*	*6.98%*	*6.05%*	*10.37%*	100.00%

Fig. 3.15 Royal Mechanism-technology matrix 9

mechanism on the rear wheel hub, the main problems which can arise from fitting disc brakes, if adopted, are that they cannot be of the same diameter as for mountain bikes or have a sporty look. The "new paintwork process" is derived from tube painting in the thermo-technical industry.

Now, go back to QFD.

We create the Mechanisms-Technology matrix 9 (Fig. 3.15), to weight how applicable new technology is to each specific mechanism. In this case the priority of mechanisms, which keeps trace of the Functions priority and the Customer Requests priority, is used as an input for the calculation of the Technology priority.

You see that the team is led very accurately towards the technical configuration of the new product.

In the Royal bicycle case study we see that "Rear hub gear", "Disc brakes" and "Light frame", obtain the highest scores (Fig. 3.16).

Exercise Let's create your Technologies Deployment Chart and Technologies matrix 9. Take the Royal case study as a guideline. Where possible, set up the formulas needed to automate computation in the software spreadsheet. Follow these stages.

(a) Create the Technologies Deployment Chart; look for technologies with an open mind and sense of reality. Forget technologies which are impossible to reach with your budget. Take a day to think about this problem and use your creativity. Find a maximum of 10 technologies.

(b) Create Mechanisms—Technologies matrix 9.

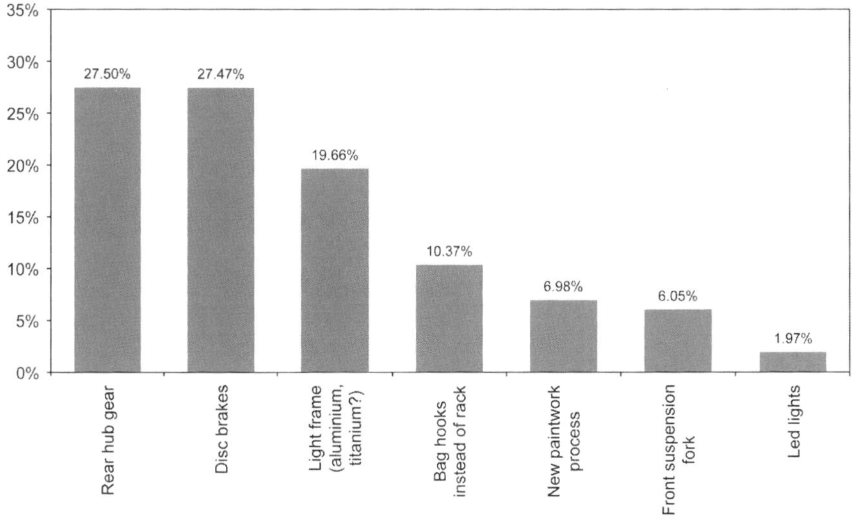

Fig. 3.16 Technology priority, Royal bicycle

(c) Technologies priorities should be calculated automatically.
(d) Create the histogram and comment on it.

Take note of the time spent and energy used during this phase.

3.6 Parts Deployment, Costs and Process Deployment

The technical solutions we presented will influence the "Parts" that make up the new product. We will calculate their importance and costs, targeted or quoted, and at the same time identify the bottlenecks cost (Fig. 3.17).

3.6.1 The Parts of the Product

The product "Parts" are its components, from a pre-production point of view.

Using data from mechanisms, from new technologies and characteristics, a KJ session will lead to Parts deployment and creating the *Parts Deployment Chart* (PDC).

If we analyse the product components in detail, the PDC might become very complex, because of the number of components. Often, to avoid over large matrices, the work is divided into sub-systems, similar to mechanisms but not

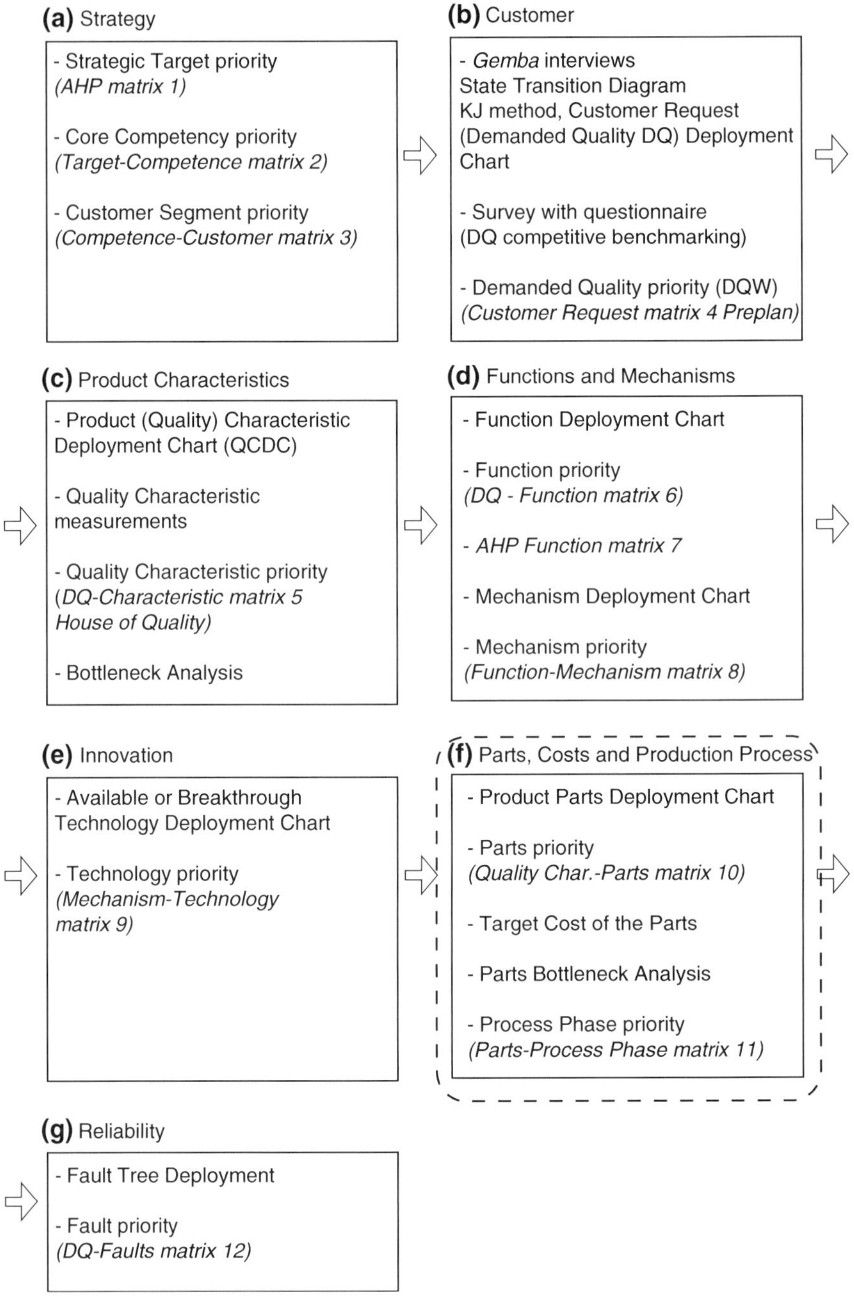
(a) Strategy
- Strategic Target priority
(AHP matrix 1)
- Core Competency priority
(Target-Competence matrix 2)
- Customer Segment priority
(Competence-Customer matrix 3)
(b) Customer
- Gemba interviews
State Transition Diagram
KJ method, Customer Request
(Demanded Quality DQ) Deployment
Chart
- Survey with questionnaire
(DQ competitive benchmarking)
- Demanded Quality priority (DQW)
(Customer Request matrix 4 Preplan)
(c) Product Characteristics
- Product (Quality) Characteristic
Deployment Chart (QCDC)
- Quality Characteristic
measurements
- Quality Characteristic priority
(DQ-Characteristic matrix 5
House of Quality)
- Bottleneck Analysis
(d) Functions and Mechanisms
- Function Deployment Chart
- Function priority
(DQ - Function matrix 6)
- AHP Function matrix 7
- Mechanism Deployment Chart
- Mechanism priority
(Function-Mechanism matrix 8)
(e) Innovation
- Available or Breakthrough
Technology Deployment Chart
- Technology priority
(Mechanism-Technology
matrix 9)
(f) Parts, Costs and Production Process
- Product Parts Deployment Chart
- Parts priority
(Quality Char.-Parts matrix 10)
- Target Cost of the Parts
- Parts Bottleneck Analysis
- Process Phase priority
(Parts-Process Phase matrix 11)
(g) Reliability
- Fault Tree Deployment
- Fault priority
(DQ-Faults matrix 12)

Fig. 3.17 Framework phase **f**, Parts, Costs and Production process

necessarily the same. We must keep in mind that mechanisms arise from their ability to carry out functions.

In the following list we can see the Royal bicycle project PDC.

- Handlebars;
- Led lights;
- Brake system and levers;
- Saddle;
- Aluminium frame;
- Hub gear and chain;
- Chain-guard and mudguards;
- Rack;
- Wheels.

We notice that some innovations were adopted after the mechanisms and technologies analysis was made.

We can now create the matrix and calculate the Parts priority.

In this case, it is better to use the technical characteristics as an input (see Fig. 3.2); in fact the parts might be assessed from an engineering point of view. The voice of the customer has already influenced the priority of functions, mechanisms and technologies, which are essential to make the product unique.

The QFD team creates the *Characteristic—Parts matrix 10* (sub-systems) and fills it in, according to the impact that the Parts have on the Quality Characteristics. This matrix is similar to the four matrices model as used by Hauser and Clausing (1988, see Chap. 1). In Fig. 3.18 you can see matrix 10 for the Royal bicycle project. Parts priority has the usual decreasing exponential trend. "Aluminium frame" and "Chain-guard and mudguards" represent the most important parts for the new product, from a technical point of view (Fig. 3.19).

Exercise Let's create your Parts Deployment Chart and Parts matrix 10. Take the Royal case study as a guideline. Where possible, set up the formulas needed to automate calculation in the software spreadsheet. Follow these stages.

(a) Create the Parts Deployment Chart. Focus on sub-systems. Find a maximum of 10 parts.
(b) Create Quality Characteristic—Parts matrix 10.
(c) Parts priorities should be calculated automatically.
(d) Create the histogram and comment on it.

Take note of the time spent and energy used during this phase.

CHARACTERISTICS - PARTS (SUBSYSTEMS) MATRIX 10											
		PARTS DEPLOYMENT									
QUALITY (PRODUCT) CHARACTERISTICS DEPLOYMENT	Product characteristic priority	Handlebars	Led lights	Disc brake system and levers	Saddle	Aluminium frame	Hub gear and chain	Chain-guard and mudguards	Rack	Wheel	
Braking noise level	8.28%			⊙						⊙	
Chain noise level	9.04%						⊙	⊘			
Hub noise level	7.37%						⊙			⊙	
Bicycle weight	5.25%	∴	∴			⊙		⊘	⊘	⊙	
Available colours	5.25%				∴	⊙		⊘	∴	∴	
Accessories	14.89%	∴	⊘		∴			∴	⊙		
Component labelling	16.10%	⊙			⊙	⊙		⊙		∴	
Bike setting	5.36%		∴	⊙	∴		⊙			∴	
Paintwork durability test	6.34%	∴				⊙		⊙	⊘		
Warranty	11.86%	∴	∴	⊘	∴	∴	⊘	∴	∴	∴	
Product delivery time	1.09%			⊘		⊘		⊘			
Lighting efficiency	2.28%		⊙								
Braking distance	6.89%			⊙						∴	Total
	Absolute weight	1.71	0.88	2.24	1.82	3.12	2.31	2.91	1.86	2.34	19.18
	PARTS PRIORITY	*8.94%*	*4.57%*	*11.66%*	*9.50%*	*16.25%*	*12.07%*	*15.15%*	*9.69%*	*12.17%*	100.00%
	Parts target cost	**€ 53.62**	**€ 27.41**	**€ 69.95**	**€ 57.02**	**€ 97.49**	**€ 72.40**	**€ 90.92**	**€ 58.15**	**€ 73.05**	€ 600.00
	Parts valued cost	**€ 45.00**	**€ 25.00**	**€ 90.00**	**€ 50.00**	**€ 130.00**	**€ 55.00**	**€ 50.00**	**€ 60.00**	**€ 95.00**	€ 600.00
	Delta	**€ 8.62**	**€ 2.41**	-€ 20.05	**€ 7.02**	-€ 32.51	**€ 17.40**	**€ 40.92**	-€ 1.85	-€ 21.95	**€ 0.00**

Fig. 3.18 Royal Characteristic—Parts matrix 10

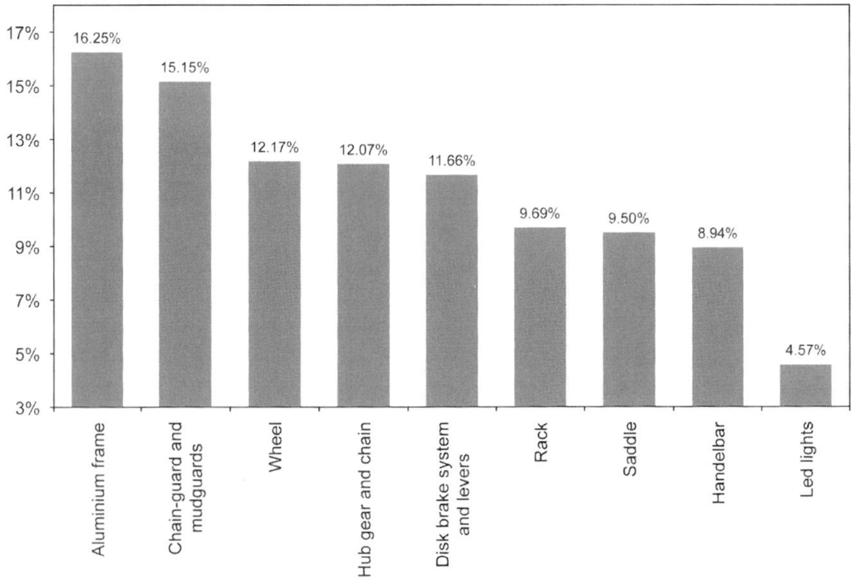

Fig. 3.19 Product Parts priority, Royal bicycle

3.6.2 The Costs of the Parts

At this stage we can introduce some product cost analysis because our technical choices can create problems for the costs of the new product.

Maekawa and Ohta (in Akao 1990) note that new product development must meet issues of cost, quality and lead time.

We have to answer the questions:

- How can the cost of a product be subdivided?
- What costs arise from the parts of the product being produced in house or purchased?

Let's consider a target price, determined by analysing the competitors and the market requests. For the new Royal bicycle a target price is set at 1,300–1,400 Euros.

From the price we can calculate a "*target cost*", thanks to internal assessments (when possible using a benchmarking on the product profit competitors have). For the new Royal bicycle the target cost is set at 600 Euros.

Usually the target cost is divided according to the value of each component as indicated by suppliers or the production department. (*Parts value*). Here we can refer to the previous version of the product and use the information from the bill for materials and parts (and their costs).

An alternative way, and maybe even a revolutionary one, is to:

divide the target costs according to the priority of the parts (Parts Target Cost).

The same evaluation could be carried out for mechanisms, functions or available technologies.

At the bottom of matrix 10 (Fig. 3.18) Parts Target Costs and Parts value are shown; the known part value is calculated by the pre-production department, or it is the purchase cost from the supplier, depending on the make or purchase policy.

In the Royal case study all the costs are from supplier purchases, nothing is manufactured inside the company. These costs include assembly and fine tuning in the production chain.

Let's compare the two types of costs.

Parts value can be higher, lower or similar to the target cost we calculated with the relative weights of the parts.

The difference δ (Delta) is given by:

$$\delta = \text{Part target cost} - \text{Part value}$$

If $\delta > 0$ it is therefore possible to invest more money for the improvement of the Part. In this case technologies or customer requests are verified again, to decide if the improvement will be from a technical and/or a styling point of view.

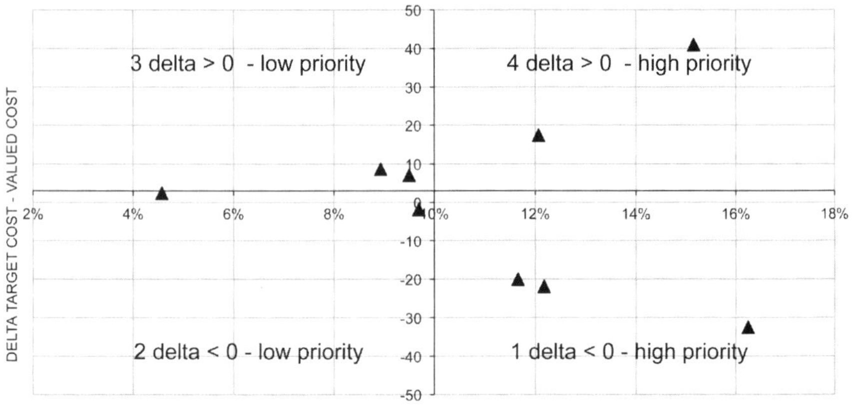

Fig. 3.20 Cost parts bottleneck analysis, Royal bicycle

If $\delta < 0$ the supplier/production cost is higher than the target cost; this is a bottleneck when the component has a high relative weight.

We can already guess the following phase.

Coming back to the bottleneck characteristics analysis, it is possible to create a similar diagram, placing δ value on the Y axis and the priority of the parts on the X axis (Fig. 3.20).

Let's analyse the four sections of the diagram:

1. Section 1 includes high importance components and negative δ; they are *bottlenecks*, the parts values are too high and a way to reduce them needs to be found. "Disc brake system and levers", "Aluminium frame" and "Wheels" are bottlenecks for new Royal.
2. Section 2 includes low importance parts with a negative δ; the production/supplier costs are higher than the target costs; here we have to reduce the part costs, as the importance of these is relatively low.
3. Section 3 concerns low importance parts with positive δ.
4. This section concerns high importance components and a positive cost gap; the parts values are lower than the target costs. It is clear that we can invest more money for these sub-systems, increasing purchase costs and/or production costs; the dots in this section will move towards the X axis.

Exercise Now add 3 rows at the bottom of your House of Quality: Parts target costs, Parts value and Delta. Then with the help of your spreadsheet software create a graph similar to Fig. 3.20. Study the deployment of the points. What are the bottlenecks in your project? What parts are located in area No. 4?

3.6.3 Process Phase Priority

This is the production process phase of our product development planning.

It is important that you remember that our team is still working through the first phases of the new product development process; it is still only on paper.

In other words the framework does not describe what has actually happened "during" the NPD process but is helping the team to "plan" during the early stages of the development process (see Chap. 1).

The same can be said about the planning for the production phase. We plan these phases according to Parts priority, which comes from technologies, mechanisms, functions and characteristics.

This process allows for the product parts to be made and/or assembled in the final complete product. In the Royal case study we will consider final assembly. The importance of these phases can be a good indicator for any necessary investments. The same approach we used for costs of the parts can be carried out for the process phases.

A *Phase Deployment Chart* for the Royal bicycle is presented in the list below which follows a typical assembly sequence for a bike.

- Graphics;
- Fork insertion on frame and handlebar;
- Hub gear on frame assembly;
- Cable Housing;
- Fitting of Brakes;
- Wheel assembly on frame;
- Fitting of Chain-guard and mudguards;
- Fitting of Lights;
- Saddle and accessories.

Figure 3.21 shows Parts—Process phase matrix 11 (Royal bike). Filling the matrix takes place as usual and shows how the Process phases are correlated to product Parts.

Matrix 11 output is the Parts priority shown in Fig. 3.22.

Surprisingly for the team, "Graphics" is the most important process phase. A link between customer and production is evident. "Cable housing" and "Fitting of Brakes" are positioned quite high in priority, showing, together with "Hub gear on frame assembly", how important the disc brakes and hub gear (two of the main innovations adopted for this bicycle) assembly phases are.

We know that "Disc brake system and levers", "Aluminium frame" and "Wheels" are parts cost bottlenecks for the new Royal, but the situation is different between the former and the latter two.

For "Aluminium frame" and "Wheels" the problem lies in the choice of supplier. This choice needs to be made with the target cost in mind, in other words the parts value in matrix 10 indicates that a new supplier survey is needed.

PARTS - PROCESS PHASE MATRIX 11											
		PROCESS PHASE DEPLOYMENT									
PARTS DEPLOYMENT	Parts priority	Graphics	Fork insertion on frame and handlebar	Hub gear on frame assembly	Cable Housing	Fitting of brakes	Wheel assembly on frame	Fitting of Chain-guard and mudguards	Fitting of lights	Saddle and accessories	
Handlebars	8.94%	⊘	⊙		⊘	⊘				⁂	
Led lights	4.57%								⊙		
Disc brake system and levers	11.66%				⊙	⊙	⁂				
Saddle	9.50%	⁂								⊙	
Aluminium frame	16.25%	⊙	⊙		⊘		⁂	⊘	⊘		
Hub gear and chain	12.07%			⊙	⊙		⁂				
Chain-guard and mudguards	15.15%	⊙		⊘				⊙	⊘		
Rack	9.69%	⁂							⊙	⊙	
Wheel	12.17%	⊘		⊙		⊙	⊙				Total
	Absolute weight	3.65	2.27	2.64	2.89	2.41	1.50	1.85	2.23	1.82	21.25
	PARTS PRIORITY	*17.19%*	*10.67%*	*12.41%*	*13.61%*	*11.36%*	*7.04%*	*8.71%*	*10.47%*	*8.55%*	100.00%

Fig. 3.21 Royal Parts—Process phase matrix 11

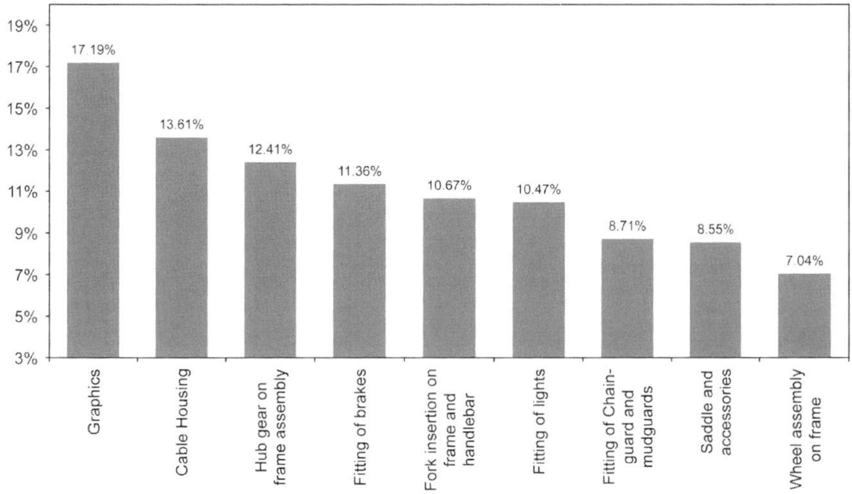

Fig. 3.22 Process Phase priority, Royal bicycle

On the other hand, new suppliers need to be chosen for the "Disc brake system and levers" as well as careful planning as regards assembling because of the components of this sub-system.

Exercise Let's create your Process phase Deployment Chart and the Process phase matrix 11. Take the Royal case study as a guideline. Where possible, set up the formulas needed to automate computation in the software spreadsheet. Follow these stages.

(a) Create the Phase Deployment Chart. Ask production managers to help your team. Remember that in the case of service development, production phases correspond to service delivery to customer.
(b) Create Parts—Process phase matrix 11.
(c) Phase priorities should be calculated automatically.
(d) Create the histogram and comment on it.

Take note of the time spent and energy used during this phase.

3.7 Reliability

QFD also deals with reliability of the product or service. Our Framework ends here (Fig. 3.23), with phase g).

As shown in Fig. 3.2, from the point of view of the workflow, Reliability deployment and calculation of Faults priority depend on Parts planning and its effects on customer satisfaction and therefore on Customer Requests (Demanded Qualities).

It is like coming back in a loop of continued quality improvement, to customer expectations.

First of all the working team must build a *Fault Deployment Chart*, that we can also call "Fault Tree". It shows all the faults and defects that can occur, and organises them in a hierarchical structure. Sticky notes using KJ and a close relationship with the Quality Department help identify Faults.

In the Royal bicycle case, the Fault Tree is shown in Table 3.2.

We now have to weight the product faults, considering their impact on satisfaction of customer requests.

Here we use the *Request-Fault matrix 12*, with the Demanded Quality DQ in the rows and the Fault Tree FT in the columns. Thanks to the Independent Scoring Method, filling this matrix in leads to absolute values and then to relative values, the Fault priority (Figs. 3.24 and 3.25).

If we analyse the Fault priority histogram we notice that the formation of "rust" is particularly important; the team developing the product will have to pay great attention to durability and resistance tests (carried out in saline mist) of the paintwork, the finishing of the chassis, fork and other components.

Faults concerning "Jerky cycling" or "Intermittent braking" follow closely, indicating why prototype testing is very important.

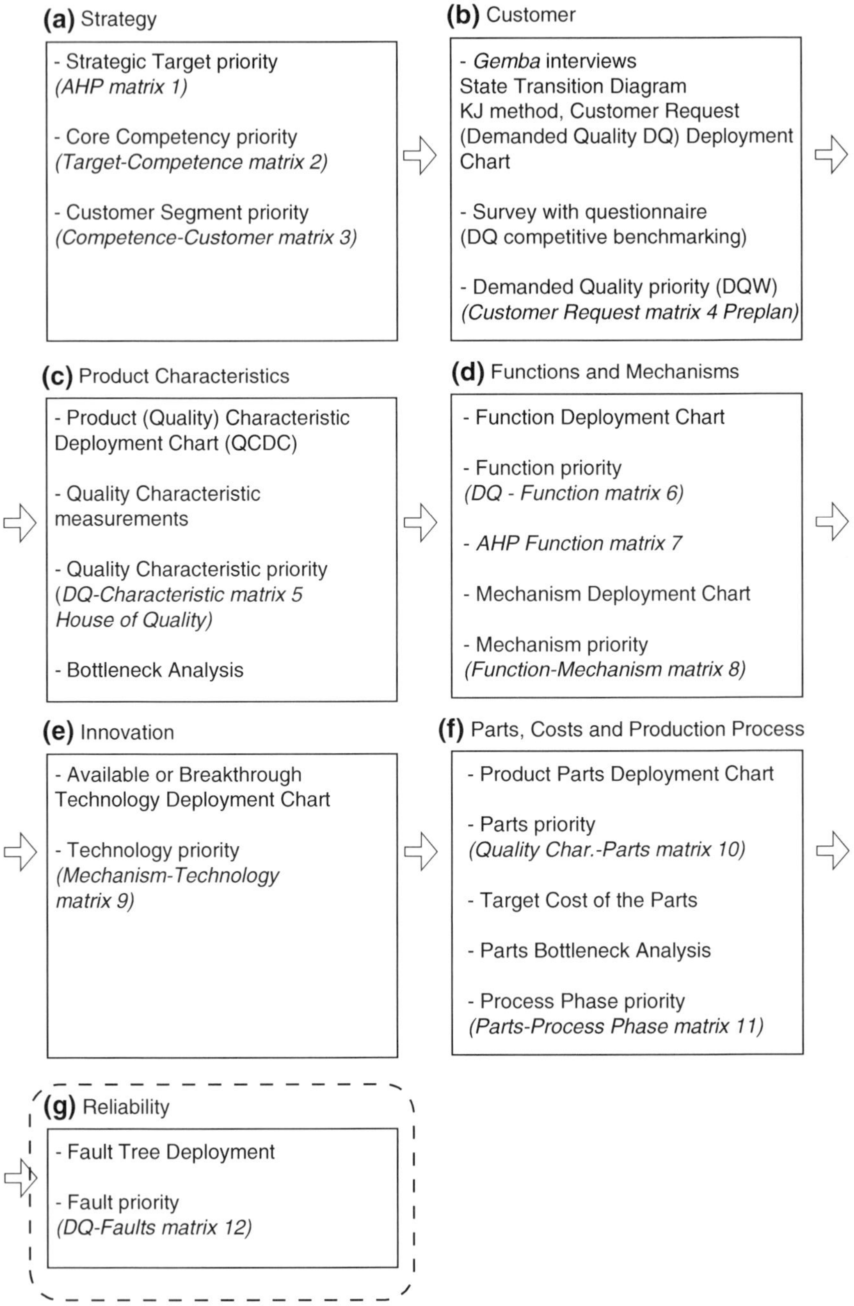
(a) Strategy
- Strategic Target priority
(AHP matrix 1)
- Core Competency priority
(Target-Competence matrix 2)
- Customer Segment priority
(Competence-Customer matrix 3)
(b) Customer
- Gemba interviews
State Transition Diagram
KJ method, Customer Request
(Demanded Quality DQ) Deployment
Chart
- Survey with questionnaire
(DQ competitive benchmarking)
- Demanded Quality priority (DQW)
(Customer Request matrix 4 Preplan)
(c) Product Characteristics
- Product (Quality) Characteristic
Deployment Chart (QCDC)
- Quality Characteristic
measurements
- Quality Characteristic priority
(DQ-Characteristic matrix 5
House of Quality)
- Bottleneck Analysis
(d) Functions and Mechanisms
- Function Deployment Chart
- Function priority
(DQ - Function matrix 6)
- AHP Function matrix 7
- Mechanism Deployment Chart
- Mechanism priority
(Function-Mechanism matrix 8)
(e) Innovation
- Available or Breakthrough
Technology Deployment Chart
- Technology priority
(Mechanism-Technology
matrix 9)
(f) Parts, Costs and Production Process
- Product Parts Deployment Chart
- Parts priority
(Quality Char.-Parts matrix 10)
- Target Cost of the Parts
- Parts Bottleneck Analysis
- Process Phase priority
(Parts-Process Phase matrix 11)
(g) Reliability
- Fault Tree Deployment
- Fault priority
(DQ-Faults matrix 12)

Fig. 3.23 Framework phase **g**, Reliability

Table 3.2 Fault Deployment Chart, Royal bicycle

It does not move	Jerky cycling
	Blocked pedals
It lights up the road ahead badly	No light
	Intermittent light
It does not brake	Intermittent braking
	Braking not fluid
Rust	Rust

CUSTOMER REQUEST - PRODUCT FAULT MATRIX N. 12									
		FAULT DEPLOYMENT (TREE)							
		It does not move		It lights up the road badly		It does not brake		Rust	
CUSTOMER REQUEST DEPLOYMENT	Customer request priority	Jerky cycling	Blocked pedals	No light	Intermittent light	Intermittent braking	Braking not fluid	Rust	
Feeling of refinement and classic style	11.16%			⁂	⊘			⊙	
Saving time in assembling	8.82%	⁂				⁂	⁂		
Riding a status symbol	9.92%				⊘		⊘	⊙	
Make an investment for the future	12.54%							⊙	
Absence of back stress	5.16%	⁂				⁂			
Regular progression in braking	12.27%					⊙	⊘		
Feeling of silence while riding	7.95%	⊙	⁂						
Feeling of fluidity while pedalling	6.78%	⊙	⊘						
Need for sports style	4.82%					⁂	⁂	⁂	
I feel different from the other cyclists	11.87%							⁂	
Feeling of craftsmanship	8.68%	⊘	⊘					⊘	Total
	Absolute weight	1.73	0.54	0.11	0.63	1.29	0.80	3.45	8.56
	Faults priority	*20.16%*	*6.35%*	*1.30%*	*7.39%*	*15.09%*	*9.37%*	*40.33%*	100.00%

Fig. 3.24 Royal Customer Request-fault matrix 12

Exercise This is the last one. Focus on Faults. Ask for the help of the Quality Department. Focus firstly on a small number of defects or faults that can affect the customer. Later you could use the same approach for production process faults. Take the Royal case study as a guideline. Where possible, set up the formulas needed to automate calculation in the software spreadsheet. Follow these stages.

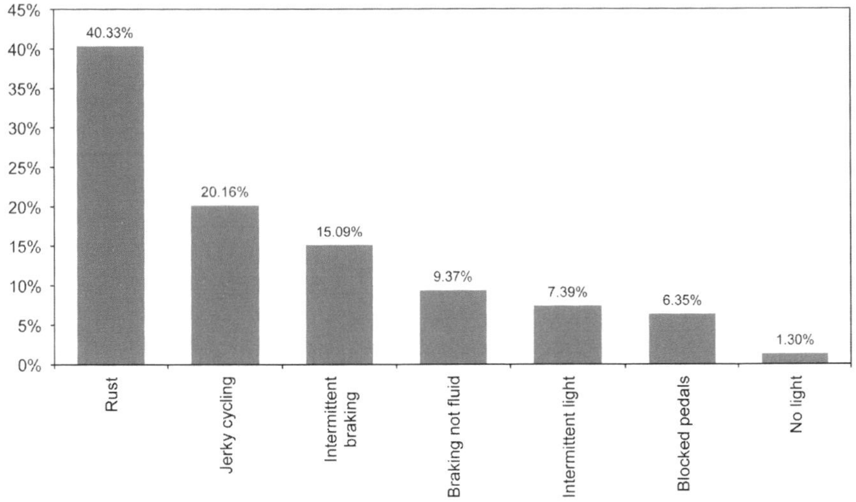

Fig. 3.25 Royal Fault priority

(a) Create the Fault Deployment Chart. Ask quality experts to help your team.
(b) Create Customer Request—Fault matrix 12.
(c) Fault priorities should be calculated automatically.
(d) Create the histogram and comment on it.

Take note of the time spent and energy used during this phase.

3.8 Summary

In this chapter Quality Function Deployment demonstrates its potential for technological product planning.

Apart from House of Quality it is possible to draw a practical framework for technicians up to the production stage.

We start from the identification of the product/service characteristics with the purpose of creating the Request-Characteristic matrix 5 (House of Quality). The technical benchmarking follows, in order to highlight the bottleneck characteristics.

After characteristics, Functions are identified, and the Customer Request-Function matrix and AHP matrices 6 and 7 (Functions technical point of view) are created.

From the Functions we deploy Mechanisms, macro-systems which carry out the Functions. Functions-Mechanisms matrix 8 allows us to define the importance of Mechanisms and from them the Technology importance through the link Mechanism-technology matrix 9.

Technologies, together with Mechanisms, allow us to identify Product Parts. The Characteristic-Parts matrix 10 output is the Parts priority, from this weight it is possible to give a new allocation to the "target costs"; an analysis of the part bottlenecks is also possible.

From Parts it is just a short step to Process phases: a Parts-Process phase matrix 11 is used to calculate Process phase priority.

Finally, Reliability is strictly linked to Parts and Customer Request thanks to the Fault Tree. Fault priority allows the company's Quality Department to take corrective action to reduce the probability and effects of faults starting from the most important ones.

References

Akao Y (ed) (1990) Quality function deployment QFD, integrating customer requirements into product design. Productivity Press, Cambridge

ASI American Supplier Institute (1987) Quality function deployment. Instruction manual. ASI, Dearborn

Hauser JR, Clausing D (1988) The house of quality. Harvard Bus Rev 66(3):63–73

Kano N, Seraku N, Takahashi N, Tsuji S (1984) Attractive quality and must-be quality. J Soc Quality Control 14:39–48

Mazur G (1999) Comprehensive quality function deployment for products. Product version 2000. QFD Network

Fuzzy QFD

4

Abstract

Fuzziness concept, i.e. vagueness, is an essential part of the human mind and philosophy, like nature and the environment, which are guided by some mathematical rules but also by vagueness. A simplified approach to fuzzy mathematics applied to QFD is proposed in this chapter. The aim is to give the reader some basic tools to be used if the team wants to manage fuzziness in QFD projects. The reader can consider, for example, how much fuzziness exists when she/he fills in a questionnaire or puts a symbol of correlation inside a QFD matrix. In other words, interviewees, QFD team people and specialists from the departments involved in a QFD project, all are humans and so, they can sometimes change their opinion or are unable to express what they really feel. Starting from "fuzzy sets" and operations with fuzzy sets, we will focus on the so called "fuzzy numbers", that are particular fuzzy sets. Then, the mathematical operations between fuzzy numbers, like addition, subtraction, product and division are presented with examples. At a later stage we will return to the traditional, or "crisp", mathematics and workflow, which the previous chapters are founded on. We will review, from a fuzzy point of view, some QFD process phases, like fuzzy questionnaire design, fuzzy Preplan matrix and fuzzy QFD matrices. We will learn how to calculate the final ranking of sentences, examined in the QFD matrices, and how to set a ranking order through "defuzzification" of fuzzy numbers.

4.1 Introduction

Fuzziness concept, i.e. vagueness, is an essential part of the human mind and philosophy, like nature and the environment, which are guided by some mathematical rules but also by vagueness. Vagueness is part of each single animal and plant and their unique and barely repeatable vital activities.

D. Maritan, *Practical Manual of Quality Function Deployment*,
DOI 10.1007/978-3-319-08521-0_4

Fuzzy mathematics helps when scientists want or have to describe and manage fuzziness in natural or artificial environments. It is a peculiar kind of mathematics that studies fuzzy sets and fuzzy logic.

For several reasons the use of fuzzy mathematics, and its practical applications, is becoming more diffuse.

Also in Quality Function Deployment projects.

You can consider, for example, how much fuzziness exists when you fill in a questionnaire or put a symbol of correlation inside a QFD matrix. In other words, interviewees, QFD team people and specialists from the departments involved in a QFD project, all are humans and so, they can change their opinion or are unable to express what they really feel.

The fuzzy mathematics approach is not without its limitations and inconsistencies. Some fuzzy logic rules are ambiguous and can be solved in different ways.

And fuzzy QFD is not simple. We developed some QFD projects using fuzzy mathematics (see Chap. 5) and often the results did not justify the amount of team work. Some errors should be avoided if you are thinking of using fuzzy QFD in a real project. We will discuss this later.

In spite of this, today, fuzzy mathematics is, undeniably, a fascinating research area, that tries to understand the vagueness of natural environments better.

It is best to read this chapter after having read chapters two and three. The reader must have some familiarity with mathematical formulae and classical logic sets theory. What we will look at in this chapter is not exhaustive of Fuzzy set theory but only a basic presentation of the tools needed for a fuzzy QFD fuzzy calculation.

4.2 What You Will Learn in This Chapter

A simplified approach to fuzzy mathematics applied to QFD is proposed in this chapter. The aim is to give to the reader some basic tools to be used if the team wants to manage fuzziness in QFD projects.

Starting from "fuzzy sets" and operations with fuzzy sets, we will focus on the so -called "fuzzy numbers", that are particular fuzzy sets.

Then, the mathematical operations between fuzzy numbers, like addition, subtraction, product and division are presented with examples.

At a later stage we will return to the traditional, or "crisp", mathematics which the previous chapters are founded on. We will review, from a fuzzy point of view, some QFD process phases, like questionnaire design, Preplan matrix 4 and QFD matrices (Fig. 4.1).

We will learn how to calculate the final ranking of sentences examined in the QFD matrices and how to set a ranking order through "defuzzification" of fuzzy numbers.

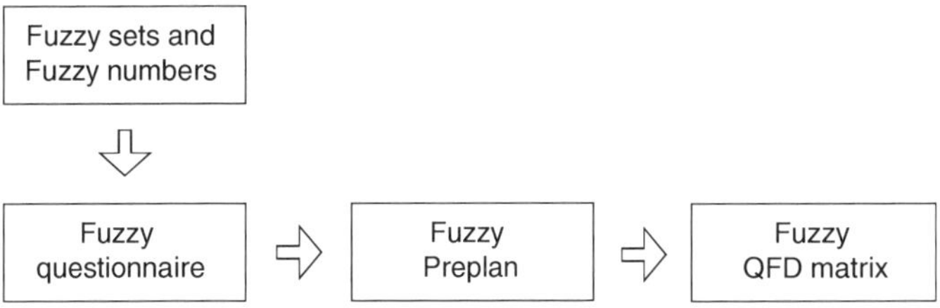

Fig. 4.1 Workflow of this chapter; from fuzzy sets and fuzzy numbers to fuzzy QFD questionnaire, Fuzzy Preplan and Fuzzy QFD matrices

4.3 Fuzzy Sets and Fuzzy Mathematics

Classical logic, introduced by Aristotle in IV century BC, is based on the "true" and "false" contrast, that is to say *bivalent.*

Let's call:

- X, the Universe set and
- A, a sub-set of X

The so called membership function $f_A(x)$ describes the state of an element x of A. in other words, in classical logic there are only two possibilities:

- x belongs ($f_A(x) = 1$) to A set or
- x does not belong ($f_A(x) = 0$) to A set

Two important principles of classical logic have to be remembered:

- Law of non contradiction (LNC);
- Law of the excluded middle

Law of non contradiction states that an element x either belongs to set A or doesn't belong to A; in this second case we say that it belongs to the "complement set" (negation) AC of A. Intersection between A and AC is the empty set.

Law of the excluded middle, also known as the law of the excluded third, states that the union of set A and its complement AC is the Universal set X. In other words every element x belongs to X.

This traditional logic, because of its nature, suffers when complex systems are analysed (Zadeh 1973). It is not able, for example, to describe the vagueness of human assessments and thoughts.

For example, concepts like "tall" and "short", "hot" and "cold", "good" and "bad", remain within classical logic with some difficulty. Several philosophers found paradoxes that are outside Aristotle's logic.

One of them is the so-called liar's paradox, attributed to the Greek philosopher Eubulides of Miletus (IV century BC) by Diogene Laerzio. He said: ψευδόμενος (*pseudòmenos*), "I lie". This sentence, if interpreted as "I, myself, at this moment, am lying", is not true then at the same time it is not false.

Aristotle himself proposed contradictory sentences (*antiphasis*) and some of them were focused on future events: is every proposition about the future either true or false? He considers these two statements:

- There will be a sea-battle tomorrow.
- There will not be a sea-battle tomorrow.

It would seem that one of them is true, the other false. But, if the first statement is now true, there must be a sea-battle tomorrow and therefore impossible that there won't be a sea-battle tomorrow. We conclude that nothing is possible except what actually happens (Smith 2014).

In 1965 Lofti A. Zadeh, professor at Berkeley University, California (Zadeh 1965), and a German mathematician Klaua (Gottwald 2010), separately but at the same time presented the so called fuzzy sets theory. Zadeh's fuzziness theory is based on the consideration that as the complexity of a system increases, our capacity to describe its behaviour and environment well decreases.

This is an attempt to deal with vagueness and uncertain situations with some scientific accuracy. Consider, for example, in human language, sentences used in questionnaires such as "rather important", "very important" "not sufficient", "very bad", "good", "very good" are not as "crisp" as words like "married", "brother", "male", "female".

As can be easily understood, fuzzy logic is particularly effective in the management of assessments and personal opinions. This is why it can be applied in some QFD projects.

What we will call "fuzzy numbers" can be interpreted as tools to calculate with words instead of classical numbers" (Zadeh 1996).

4.3.1 Fuzzy Sets

A *fuzzy set* is a set where there is a crisp border between elements that belong to the set and that do not belong. A fuzzy set extends the concept of the classical set, also called *crisp set*.

Let's focus on linguistic variables.

From a formal point of view if X = {x} is the set of all the possible values of the linguistic variable x, a fuzzy set F on x is characterized by pairs each made up of:

- A linguistic variable value;
- A corresponding *membership degree* MF(x) to F set, according to a membership function:

$$F = \{(x, MF(x) \mid x \in X, MF(x) \in [0, 1]\}$$

In other words, MF describes the "shape" of a fuzzy set and links each value of the linguistic variable to a Real number included in the [0, 1] interval (Fig. 4.2).

0 and 1 mean respectively no membership to F and full membership; other values mean partial membership of the set.

Therefore a crisp set is a fuzzy set with all the degrees of membership equal to 1.

A fuzzy set F is defined "*normalized*" if at least one element x of F has MF (x) = 1. A fuzzy set F can be normalized if it is divided by $\sup_x$ MF(x), i.e. the maximum value of the membership function.

A fuzzy set is called "*convex*" if and only if its membership function is convex (see for example Fig. 4.2)

What *operations* are possible between fuzzy sets? Here some differences exist between fuzzy sets and crisps sets. Let's call A and B two fuzzy sets.

Union (Fig. 4.3):

$$(A \cup B)(x) = max\,(A(x), B(x))$$

Intersection (Fig. 4.3):

$$(A \cap B)(x) = min\,(A(x), B(x))$$

Inclusion:

$$A \subseteq B\, se\, A(x) \leq B(x) \forall x \in X$$

Complement:

$$\bar{A}(x) = 1 - A(x)$$

Fig. 4.2 An example of a fuzzy set membership function *MF(x),* x-axis represents the fuzzy set elements (linguistic variables)

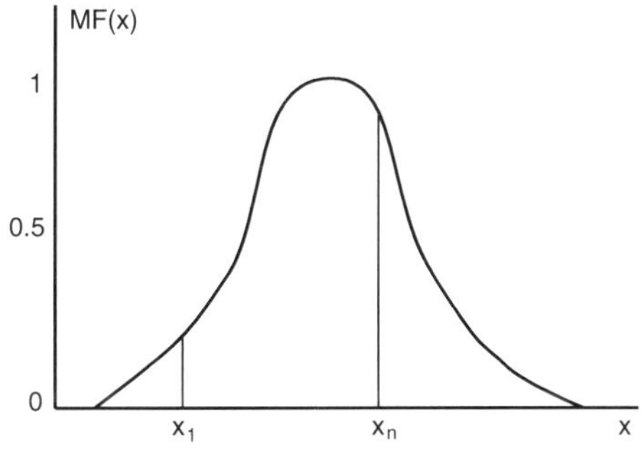

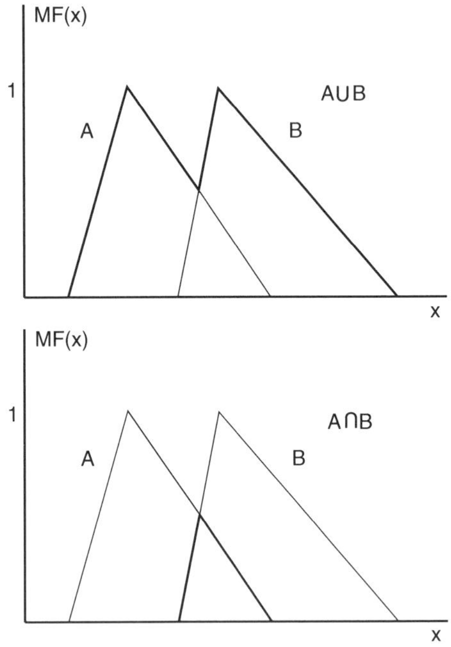

Fig. 4.3 Union and intersection of two fuzzy sets A and B. The membership function, in this example, is supposed to be triangular. X-axis represents the fuzzy set elements (linguistic variables). Y-axis represents membership functions values

To simplify, the common rules that are applicable to crisp sets can also be for fuzzy sets too, with the exception of two principles:

- Law of non contradiction;
- Law of the excluded middle.

In other words:

$$A \cup \bar{A} \neq X$$
$$A \cap \bar{A} \neq \varnothing$$

For instance, saying that the probability that a person will wear a blue dress today is 70 %, describes two different states: this person will either wear the blue dress or will not. No other alternatives are possible. We are in the probabilistic theory of classic logic.

Saying that a person has a 0.7 degree of membership to the set "tall people" (we could say the same for happy people, or poor), means to put this element inside the set with some vagueness.

The two approaches are significantly different. In fact probability measures the statistical frequency of an event, while on the other hand membership measures a deterministic but ambiguous aspect.

4.3.2 Fuzzy Numbers

We call "*fuzzy number*" a fuzzy set F, normal and convex, on the real numbers $\Re$.

A fuzzy number F is called "triangular" fuzzy number (TFN) if its membership function MF(x) is:

$$MF(x) = \begin{cases} (x-a)/(b-a), & if\, a \le x \le b \\ (c-x)/(c-b), & if\, b \le x \le c \\ 0, & otherwise \end{cases}$$

Often b = (a + c)/2. In this case the fuzzy number is called "central triangular" or "symmetrical triangular" (STFN) (Fig. 4.4).

We call "*support*" of the fuzzy number the crisp set of Real numbers with membership degree different to zero ("a" and "c" are the two limits of the support).

Now let's see how we can operate with triangular fuzzy numbers.

Let F1 = (a1, b1, c1) and F2 = (a2, b2, c2) two triangular fuzzy numbers, r a crisp Real number (Chen and Hwang 1992).

Sum:

$$F1 + F2 = (a1 + a2, b1 + b2, c1 + c2)$$

Sum can be extended to *n* triangular fuzzy numbers.

Subtraction follows a different rule that considers the results from the subtractions between the inverted limits of the support, sorted in ascending order:

$$F1 - F2 = (a1 - c2, b1 - b2, c1 - a2)$$

Multiplication and division by crisp Real number *r*:

$$\begin{aligned} rF1 &= (ra1, rb1, rc1)\, r > 0 \\ F1/r &= (a1/r, b1/r, c1/r)\, r > 0 \end{aligned}$$

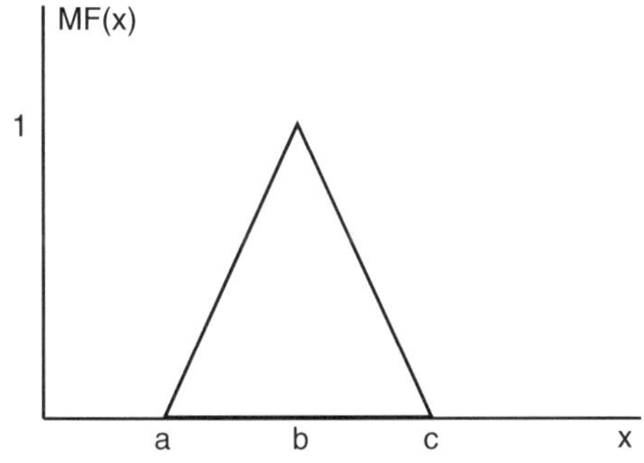

Fig. 4.4 Symmetric triangular fuzzy number (a, b, c) with b = (a + c)/2

Multiplication and division between triangular fuzzy numbers corresponds to the multiplication of the extremities (Chan et al. 1999).

$$F1 * F2 \cong (a1 * a2,\ b1 * b2,\ c1 * c2)\ a1 \geq 0,\ a2 \geq 0$$

in detail:

$$F1 * F2 = (\min(a1b1,\ a1b3,\ a3b1,\ a3b3),\ a2b2,\ \max(a1b1,\ a1b3,\ a3b1,\ a3b3))$$
$$F1/F2 \cong (a1/c2,\ b1/b2,\ c1/a2)\ a1 \geq 0,\ a2 \geq 0$$

in detail:

$$F1/F2 = (min(a1/b1,\ a1/b3,\ a3/b1,\ a3/b3),\ a2/b2,\ max(a1/b1,\ a1/b3,\ a3/b1,\ a3/b3))$$

For example: be F1 = (2, 4, 6) and F2 = (1, 2, 3) two triangular fuzzy numbers. Then:

$$\begin{aligned} \mathrm{F1} + \mathrm{F2} &= (3, 6, 9) \\ \mathrm{F1} - \mathrm{F2} &= (1, 2, 5) \\ \mathrm{F1} * \mathrm{F2} &= (2, 8, 18) \\ \mathrm{F1}/\mathrm{F2} &= (2/3, 4/2, 6/1) = (0.66, 2, 6) \\ \mathrm{F1} - \mathrm{F1} &= (-4,\ 0, 4) \\ \mathrm{F1}/\mathrm{F1} &= (2/6, 4/4, 6/2) = (0.33, 1, 3) \end{aligned}$$

This example shows that if a triangular fuzzy number is subtracted from itself, the result is not zero. Similarly, if a fuzzy number is divided by itself, the result is not one.

$$F - F \neq 0$$
$$F/F \neq 0$$

These presented rules are not completely unambiguous: some authors (Nagoor Gani and Mohamed Assarudeen 2012; Gao et al. 2009), to solve optimization problems, suggest using calculation more complex algorithms for subtraction, multiplication and division.

By and large, for QFD fuzzy matrices these approximate representations can be used.

4.4 Fuzzy QFD

In several phases of the framework presented in chapters two and three, we have to fill in a questionnaire or to assess relationships with linguistic variables.

Usually this process also requires (Zimmermann 1987):

- a final rating, i.e. a score to each element (I called it "priority" in Chaps. 2 and 3);
- a rank order of these ratings.

If these calculations seem to be quite accessible in a crisp environment, in fuzzy things become complicated. From being crisp Mathematics becomes fuzzy and this implies a new way of calculating the final rating (priority) and the rank order.

Where can fuzzy mathematics be used in QFD?

Three main areas exist:

- Fuzzy questionnaire;
- Fuzzy matrices;
- Fuzzy final rating and ranking order.

4.4.1 Fuzzy Questionnaire

The reader will remember that after KJ sessions, the Demanded Quality Deployment Chart (DQDC, framework phase "b", Chap. 2) is created. Then a survey using a questionnaire is planned.

We can link fuzzy numbers instead of crisp ones to interviewees' answers, in order to keep track of the uncertainty of answers.

In this way we could say that an answer like "not important" becomes "approximately not important".

Now let's focus on the Likert scale of the questionnaire.

In Chap. 2 we have seen that answers are ranked on a 5 point Likert scale, with the following associated statements for the first QFD question (importance assessment):

1 = not important at all
2 = a little important
3 = rather important
4 = very important
5 = extremely important

The second question concerns the assessment the interviewee gives to the ability that products/services (ours and competitors) involved in the benchmarking, have to

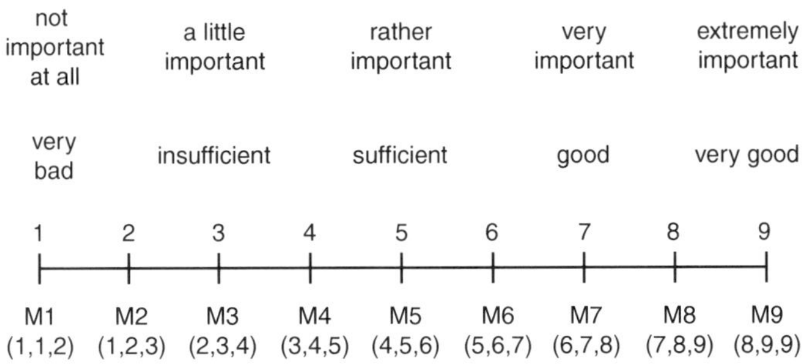

Fig. 4.5 The 9 points Likert scale with semantic expressions for answers to the two QFD questionnaire questions (importance and assessment). At the bottom you see the 9 membership functions that represent nine triangular fuzzy numbers

satisfy the Demanded Quality in question. Again, answers are ranked on a 5 point Likert scale, with the following associated statements:

1 = very bad
2 = insufficient
3 = sufficient
4 = good
5 = very good

Let's now go to the fuzzy world.

Differently to the crisp QFD approach seen in Chap. 2, we adopt a 9 point scale (Chan et al. 1999). It is simple to move from a 5 point scale to 9 points by inserting intermediate ones. Each of the 9 Likert crisp points becomes a triangular fuzzy number, represented by a group of three ordinate numbers (Chen and Hwang 1992). See Fig. 4.5.

In Fig. 4.6 the triangular membership functions are presented. Note the different "shape" of M1 and M9 in comparison to the others.

This is not the only approach for creating a fuzzy QFD questionnaire. Other authors suggest different shapes and supports.

What we can say is that if the QFD team wants to use fuzzy numbers practically, the calculation process with spreadsheets becomes a little complicated. My team and I used the approach presented above in some real projects and I can say that it works without big issues from the spreadsheet programming point of view.

How is the data that comes from questionnaires managed? We can learn a practical procedure.

- Crisp data from questionnaires are fuzzified in triangular fuzzy numbers (TFNs). For example to the answer 1 the TFN (1, 1, 2) is associated; to 6 the TFN (5, 6, 7).
- For each Demanded Quality (DQ) and for each question, averages of the vertices of the TFNs are calculated, that create the average fuzzy importance or assessment for the DQ.

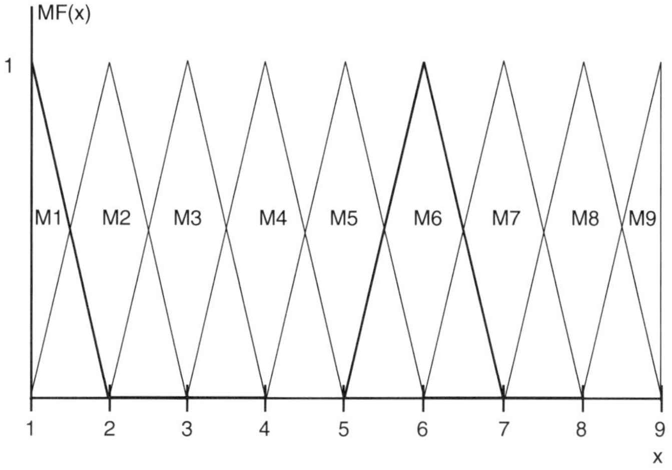

Fig. 4.6 Triangular membership functions for the 9 points answers to the QFD questionnaire; as example M1 (approximately 1) and M6 (approximately 6) are *highlighted* to show the different "shapes" for crisp points 1 and 9 and the others

We can represent the average TFN on *m* questionnaires as M′ = (a′, b′, c′) and the vertices of this triangle are built with average values:

$$a' = \frac{\sum a}{m}$$
$$b' = \frac{\sum b}{m}$$
$$c' = \frac{\sum c}{m}$$

At this point it is easy to calculate these values in a spreadsheet. The only problem is that the number of calculations is multiplied by three.

4.4.2 Fuzzy Preplan

We can now proceed with the Preplan matrix (see Chap. 2). For each Preplan column now we have to manage a group of three numbers, the average TFNs from the questionnaire.

In order to help the reader understand, let us imagine an example and work with 3 DQ and one competitor (Table 4.1). Let's see what happens in the next columns.

Quality Plan. The first problem to solve is to calculate the Quality Plan (QP), the assessment target. It is the target of perceived quality, which the development team calculates for every DQ, starting from the opinion given about our product and

Table 4.1 Example. Fuzzy average importance and Fuzzy benchmarking assessment

	Importance			Company x			Our company		
DQ1	5.8	6.7	7.5	4.3	5.0	5.8	3.3	4.4	5.2
DQ2	2.8	3.6	4.4	3.2	4.2	5.1	3.5	4.5	5.4
DQ3	6.4	7.2	8.0	5.8	6.5	7.4	5.2	6.0	6.9

about those of our competitors. In the crisp world, the logic for this area of the matrix is (Chap. 2):

- if our product has a higher average score, an assessment improvement to the specific DQ is unnecessary;
- if our product has a lower or equal to average score, an assessment improvement to the specific DQ is necessary.

With a fuzzy approach the strategy is different.

We still consider the best score, but in what way?

Consider the DQ as expressed by TFNs (a′, b′, c′) and QP as expressed by the TFN:

$$TFNQP = (a^\wedge b^\wedge c^\wedge)$$

One strategy could be identifying the vertex $b^\wedge$ of this triangle as the maximum c' value of the assessments TFNs (Fig. 4.7):

$$b^\wedge max(c')$$

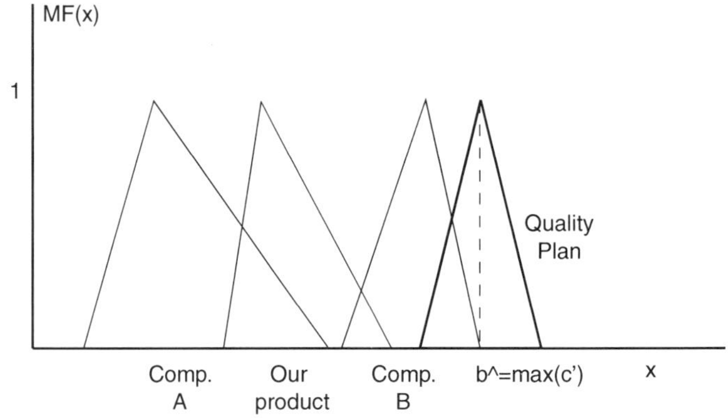

Fig. 4.7 Quality Plan Triangular membership function $TFN\ QP = (a^\wedge, b^\wedge, c^\wedge)$ with value of the vertex $b^\wedge$ as the maximum c' value of the assessments TFNs

Table 4.2 Example. TFN Quality Plan

	Importance			Company x			Our company			Quality plan		
DQ1	5.8	6.7	7.5	4.3	5.0	5.8	3.3	4.4	5.2	4.8	5.8	6.8
DQ2	2.8	3.6	4.4	3.2	4.2	5.1	3.5	4.5	5.4	4.4	5.4	6.4
DQ3	6.4	7.2	8.0	5.8	6.5	7.4	5.2	6.0	6.9	6.4	7.4	8.4

From the vertex b^ we can set a^ and b^ as:

$$a^{\wedge} = b^{\wedge} - 1$$
$$c^{\wedge} = b^{\wedge} + 1 \quad \text{if } b^{\wedge} \leq 8$$
$$c^{\wedge} = 9 \quad \text{if } b^{\wedge} > 8$$

If we consider our example, QP is calculated starting from the benchmarking between our company and company x. Results are shown in Table 4.2.

Should the assessment available only be for our product/service, TFN QP (a^, b^, c^) elements can be set as follows:

$$b^{\wedge} = c'$$
$$a^{\wedge} = a'$$
$$c^{\wedge} = c' + 1 \quad \text{if c}' \leq 8$$
$$c^{\wedge} = 9 \quad \text{if } c' > 8$$

Rate of Improvement (RI). Rate of Improvement, as in the crisp Preplan, is calculated by dividing TFN QP by the TFN average assessment for our product. In this case calculation is simple and can use the rules presented in Sect. 4.3.2.

Let's consider, for example DQ1 (Table 4.3). Fuzzy Rate of Improvement TFN RI (a, b, c) is:

$$\text{a} = 4.8/5.2$$
$$\text{b} = 5.8/4.4$$
$$\text{c} = 6.8/3.3$$

Sales Points (SP). A Demanded Quality is a sales point if it represents a key point for the company, from the point of view of brand image communication or if it is a good argument for sales.

Table 4.3 Example. TFN Rate of Improvement

	Importance			Company x			Our company			Quality plan			RI		
DQ1	5.8	6.7	7.5	4.3	5.0	5.8	3.3	4.4	5.2	4.8	5.8	6.8	0.9	1.3	2.1
DQ2	2.8	3.6	4.4	3.2	4.2	5.1	3.5	4.5	5.4	4.4	5.4	6.4	0.8	1.2	1.8
DQ3	6.4	7.2	8.0	5.8	6.5	7.4	5.2	6.0	6.9	6.4	7.4	8.4	0.9	1.2	1.6

What happens to the sales points when we move over a fuzzy calculation? We have three options (Chap. 2):

Strong sales point	TFN = (1.25, 1.5, 1.5)
Medium sales point	TFN = (1, 1.25, 1.5)
No sales point	TFN = (1, 1, 1.25)

DQ Fuzzy Weight. To calculate the fuzzy Absolute Weight (AW) we have to multiply TFN Importance (I) by TFN RI and TFN SP:

$$TFN\,AW = TFN\,I * TFN\,RI * TFN\,SP$$

We just need to use the rules presented in Sect. 4.3.2. For example (see Table 4.4):

TFN AW for DQ1 = (5.8, 6.7, 7.5) * (0.9, 1.3, 2.1) * (1, 1.25, 1.5) = (5.4, 11.0, 23.2)

Absolute weight is then normalised, i.e. TFN support is divided by the maximum value of all AW membership functions. We obtain the TFN Normalised Weight (TFN NW).

For our exercise this maximum is 23.2 (see Tables 4.4 and 4.5).

Fuzzy DQ priority. Now our objective is to put these normalized weights in order, to find a priority of the DQs.

There are several ways to sort fuzzy numbers in order, most of the times with *defuzzification methods*.

It is interesting to present the *method of* Chen (1985) and Jain (1977), who introduced the concepts of maximizing set and minimizing set to decide the ordering value of TFNs. This approach is also suggested by Chan et al. (1999) and was adopted in most of our fuzzy QFD projects.

Table 4.4 TFN Absolute Weight

	Importance			....	RI			SP	Absolute weight		
DQ1	5.8	6.7	7.5	....	0.9	1.3	2.1	⊘	5.4	11.0	23.2
DQ2	2.8	3.6	4.4	....	0.8	1.2	1.8		2.3	4.3	10.1
DQ3	6.4	7.2	8.0	....	0.9	1.2	1.6	⊙	7.4	13.3	19.4

Table 4.5 TFN Absolute Weight (TFN AW) and Normalised Weight (TFN NW)

	Absolute weight			Normalised weight		
DQ1	5.4	11.0	23.2	0.23	0.48	1.00
DQ2	2.3	4.3	10.1	0.10	0.19	0.43
DQ3	7.4	13.3	19.4	0.32	0.57	0.87

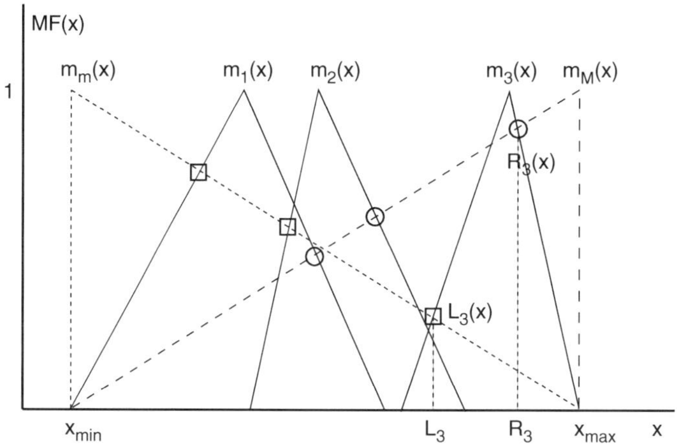

Fig. 4.8 Maximising set $m_M(x)$, *dotted right triangular* on the *right* side, and minimising set $m_m(x)$ *dotted right triangular* on the *left* side, created starting from three Triangular Fuzzy Numbers. The *circles* identify intersections with maximizing set (*Right* index $R_i(x)$), the *squares* those with minimizing set (*Left* index $L_i(x)$)

This method is centred on the intersection between fuzzy sets.

Consider n fuzzy sets $\tilde{A}_i$ (i = 1,…n), with membership functions $m_i(x)$. We define two fuzzy sets (see Fig. 4.8, the two right-angled triangles on the left and on the right):

- The maximising set M(x) with membership function $m_M(x)$;
- The minimising set N(x) with membership function $m_m(x)$.

Maximising set membership function $m_M(x)$ is:

$$m_M(x) = \begin{cases} \left[\frac{W_i(x - x_{\min})}{x_{\max} - x_{\min}}\right]^r & for\ x_{\min} < x < x_{\max} \\ 0 & otherwise \end{cases}$$

where

$$w_i = \sup(m_i(x)) \quad i = 1, \ldots n$$

When the sets are fuzzy numbers, they are normalized, then

$$\begin{aligned} w_i &= 1 \\ x_{\min} &= \min\{x \mid m_i(x) > 0; i = 1, \ldots n\} \\ x_{\max} &= \max\{x \mid m_i(x) > 0; i = 1, \ldots n\} \end{aligned}$$

As regards exponent r, it can assume different values depending on appraisal:

- Neutral (r = 1);
- Optimistic (r = 0.5);
- Pessimistic (r = 2).

We use the value r = 1 in all the calculations.

Definitely then, the membership function of the maximising set is a right-angled triangle, with the right angle on the right, height 1 and base (support) that depends on the support of the fuzzy numbers that have to be sorted in order (Fig. 4.8).

In our simple example the maximising TFN, calculated considering the normalised weights of the three DQs (Table 4.5), is the fuzzy number:

$$\mathrm{m_M}(\mathrm{x}) = (0.10, 1, 1)$$

A maximising set definition, "on the right", was worked out by Jain (1977). Chen (1985) introduces the concept of the minimizing set, in order to consider the "left" side of the membership function of the TFN to be sorted in order.

Minimising set membership function $\mathrm{m_m}(\mathrm{x})$ is:

$$\mathrm{m_m}\ (\mathrm{x}) = \begin{cases} \left[\frac{\mathrm{W_i}(x - x_{\max})}{x_{\min} - x_{\max}}\right]^r & \textit{for } x_{\min} < x < x_{\max} \\ 0 & \textit{otherwise} \end{cases}$$

where

$$w_i = \sup(m_i(x)) \quad i = 1, \ldots n$$

Again, when the sets are fuzzy numbers, they are normalized, then

$$\begin{aligned} w_i &= 1 \\ x_{\min} &= \min\{x \mid m_i(x) > 0; i = 1, \ldots n\} \\ x_{\max} &= \max\{x \mid m_i(x) > 0; i = 1, \ldots n\} \end{aligned}$$

Also here exponent "r" is equal to 1.

In our example the minimising TFN, calculated considering the normalised weights of the three DQs (Table 4.5), is the fuzzy number:

$$\mathrm{m_m}(\mathrm{x}) = (0.10, 0.10, 1)$$

Now, we need to learn the last step.

We have just seen what the maximising set "on the right" and the minimising set "on the left" are.

In general, for each fuzzy number $\tilde{A}_i$ (in our case for each TFN NW_i) we call:

- "*right index*" $R_i(x)$, the x value corresponding to the intersection between the membership function $m_i(x)$ *right side* and the maximizing set $m_M(x)$ (Fig. 4.8, intersections identified by circles), i.e.

$$R_i(x) = \min\{(m_i(x), m_M(x)\}$$

- "*left index*" $L_i(x)$, the x value corresponding to the intersection between the membership function $m_i(x)$ *left side* and the minimising set $m_m(x)$ (Fig. 4.8, intersections identified by squares), i.e.

$$L_i(x) = \min\{(m_i(x), m_m(x)\}$$

Obviously in the case of a TFN there is only one point of intersection on the right side and one on the left side of the triangle.

Finally, we calculate an index, which can sort the TFNs in order. We call it "*priority index*" ($P_i(x)$), as in the crisp environment. It has the same function.

$$P_i(x) = (R_i(x) + 1 - L_i(x))/2 \qquad \textit{for } i = 1, \ldots n$$

Now we are ready to return to our example. In Table 4.6 Normalised Weights of the three Demanded Qualities are shown.

First we have to remember that the two functions that describe the inclined sides (respectively left and right) of the generic TFN (a, b, c) are:

$$\begin{array}{ll} y = (x - a)/(b - a) & \text{if } a \leq x \leq b \\ y = (c - x)/(c - b) & \text{if } b \leq x \leq c \\ 0 & \text{otherwise} \end{array}$$

In our example the maximising set inclined side is:

$$m_M(x) = (x - 0.10)/(1 - 0.10) \quad 0.10 \leq x \leq 1$$

i.e.:

$$m_M(x) = (x - 0.10)/0.90 \quad 0.10 \leq x \leq 1$$

Table 4.6 TFN Normalised Weight

	Normalised weight		
DQ1	0.23	0.48	1.00
DQ2	0.10	0.19	0.43
DQ3	0.32	0.57	0.87

The minimizing set inclined side is:

$$m_m(x) = (x - 1)/(0.10 - 1) \quad 0.10 \leq x \leq 1$$

i.e.:

$$m_m(x) = (1 - x)/0.90 \quad 0.10 \leq x \leq 1$$

DQ1. We start with the first Demanded Quality DQ1 that has a normalized weight NW_1 = (0.23, 0.48, 1.00)

Its membership function $m_1(x)$ is (see Sect. 4.3.2):

$$MF(x) = \begin{cases} (x-a)/(b-a), & \textit{if } a \leq x \leq b \\ (c-x)/(c-b), & \textit{if } b \leq x \leq c \\ 0, & \textit{otherwise} \end{cases}$$

i.e.

$$\begin{array}{ll} y = (x - 0.23)/(0.48 - 0.23) & 0.23 \leq x \leq 0.48 \\ y = (1 - x)/(1 - 0.48) & 0.48 \leq x \leq 1.00 \\ 0, & \text{otherwise.} \end{array}$$

Let's calculate $R_1(x)$. We have to solve the system:

$$\begin{array}{ll} y = (x - 0.10)/0.90 & \text{maximising set inclined side} \\ y = (1 - x)/(1 - 0.48) & \text{triangle right side} \end{array}$$

i.e. solving in x:

$$(x - 0.10)/0.90 = (1 - x)/0.52$$

that is

$$0.58x - 0.06 = 1 - x$$

that is

$$1.58x = 1.06$$

$$x = R_1 = 0.67$$

Let's calculate $L_1(x)$. We have to solve the system:

$$\begin{array}{ll} y = (1 - x)/0.90 & \text{minimising set inclined side} \\ y = (x - 0.23)/(0.48 - 0.23) & \text{triangle left side} \end{array}$$

that is

$$(x-0.23)/0.25 = (1 - x)/0.90$$

that is

$$3.6x-0.83 = 1-x$$

That is

$$4.6x = 1.83$$
$$x = L_1 = 0.40$$

The Priority index for DQ1 is:

$$P(DQ1) = (0.67 + 1 - 0.40)/2 = 0.635$$

DQ2. DQ2 has a normalized weight $NW_2 = (0.10, 0.19, 0.43)$ Its membership function $m_2(x)$ is (see Sect. 4.3.2):

$$MF(x) = \begin{cases} (x-a)/(b-a), & \text{if } a \le x \le b \\ (c-x)/(c-b), & \text{if } b \le x \le c \\ 0, & \text{otherwise} \end{cases}$$

i.e.

$$\begin{array}{ll} y = (x - 0.10)/(0.19 - 0.10) & 0.10 \le x \le 0.19 \\ y = (0.43 - x)/(0.43 - 0.19) & 0.19 \le x \le 0.43 \\ 0, & \text{otherwise.} \end{array}$$

Let's calculate $R_2(x)$. We have to solve the system:

$$\begin{array}{ll} y = (x - 0.10)/0.90 & \text{maximising set inclined side} \\ y = (0.43 - x)/0.24 & \text{triangle right side} \end{array}$$

i.e. solving in x:

$$(x - 0.10)/0.90 = (0.43 - x)/0.24$$

that is

$$0.27x - 0.03 = 0.43 - x$$

that is

$$1.27x = 0.46$$
$$x = R_2 = 0.36$$

Let's calculate $L_2(x)$. We have to solve the system:

$y = (1 - x)/0.90$ minimising set inclined side
$y = (x - 0.10)/(0.19 - 0.10)$ triangle left side

that is

$$(x - 0.10)/0.09 = (1 - x)/0.90$$

that is

$$10x - 1 = 1 - x$$

That is

$$11x = 2$$
$$x = L_2 = 0.18$$

The Priority index for DQ1 is:

$$P(DQ2) = (0.36 + 1 - 0.18)/2 = 0.590$$

DQ3. DQ3 has a normalized weight $NW_3 = (0.32, 0.57, 0.87)$
Its membership function $m_2(x)$ is (see Sect. 4.3.2):

$$MF(x) = \begin{cases} (x-a)/(b-a), & \text{if } a \le x \le b \\ (c-x)/(c-b), & \text{if } b \le x \le c \\ 0, & \text{otherwise} \end{cases}$$

i.e.

$y = (x - 0.32)/(0.57 - 0.32)$ $0.32 \le x \le 0.57$
$y = (0.87 - x)/(0.87 - 0.57)$ $0.57 \le x \le 0.87$
0, otherwise

Let's calculate $R_3(x)$. We have to solve the system:

$y = (x - 0.10)/0.90$ maximising set inclined side
$y = (0.87 - x)/0.30$ triangle right side

i.e. solving in x:

$$(x - 0.10)/0.90 = (0.87 - x)/0.30$$

that is

$$0.33x - 0.03 = 0.87 - x$$

that is

$$1.33x = 0.90$$
$$x = R_3 = 0.68$$

Let's calculate $L_3(x)$. We have to solve the system:

$y = (1 - x)/0.90$	minimising set inclined side
$y = (x - 0.32)/(0.57 - 0.32)$	triangle left side

that is

$$(x - 0.32)/0.25 = (1 - x)/0.90$$

that is

$$3.6x - 1.15 = 1 - x$$

That is

$$4.6x = 2.15$$
$$x = L_3 = 0.47$$

The Priority index for DQ3 is:

$$P(DQ3) = (0.68 + 1 - 0.47)/2 = 0.605$$

Now we can sort the Demanded Qualities in order, following the priority indexes:

P(DQ1) = 0.635
P(DQ3) = 0.605
P(DQ2) = 0.590

As a final note, the present method is only one of the numerous methods that aim to prioritise a sequence of fuzzy numbers. Further research is definitely advisable. For example, the Chen method sometimes seems not to satisfy the transitive law

(Lotan and Koutsopoulos 1993), although we tested the results for this problem in some of our projects without finding this difficulty.

This approach can be programmed, albeit with some difficulty, using a computer spreadsheet and in several projects it proved to be sufficiently robust to support us in decision making.

If the team has time, a second defuzzification can be used to compare results.

Another simpler way for defuzzification should be added to our discussion.

It is the so called "*Medium of Maxima*" (MOM) method, which calculates the average of the x crisp values that correspond to the maxima (one or more) of the membership function:

$$MOM = \sum_{j=1}^{l} \frac{z_j}{l}$$

In case of TFNs only a maximum exists and defuzzification is very easy.

For example, the MOM defuzzified crisp value of TFN (2, 3, 5) is 3.

4.4.3 QFD Matrices and House of Quality

Classic QFD matrices, which use rows/columns correlation, expressed in symbols, and Independent Scoring Method (ISM), can be managed with fuzzy mathematics. It takes the vagueness of assessments and decisions of the QFD team into consideration.

All the tools we have seen in the previous pages can be used in these matrices. We have to add the fuzzy numbers associated to the symbols used (see Chaps. 2 and 3),

On the crisp side symbols weights are:

⊙ = strong correlation, 9
⦸ = medium correlation, 3
⁂ = weak correlation, 1
Empty cell = no correlation, 0

A possible proposal for the fuzzy world, which was used in some of our projects, is:

⊙ = strong correlation, TFN (4, 9, 9)
⦸ = medium correlation, TFN (1, 3, 5)
⁂ = weak correlation, TFN (0, 1, 2)
Empty cell = no correlation, TFN (0, 0, 1)

ISM calculation process, in this case, considers that the weight that comes from the previous matrix is a TFN, that is multiplied by the TFN corresponding to the symbol. These TFNs are then added alongside the row to get the Absolute Weight that is then normalised. Priority is calculated with the maximising and minimising set approach, as we have seen in the Preplan.

References

Chan LK, Kao HK, Ng A, Wu ML (1999) Rating the importance of customer needs in quality function deployment by fuzzy and entropy methods. Int J Prod Res 37(11):2499–2518
Chen Shan-Huo (1985) Ranking fuzzy numbers with maximizing set and minimizing set. Fuzzy Sets Syst 17:113–130
Chen SJ, Hwang CL (1992) Fuzzy multiple attribute decision making: methods and applications. Springer, Berlin
Gao S, Zhang Z, Cao C (2009) Multiplication operation on fuzzy numbers. J Softw 4(4):331–338
Gottwald S (2010) An early approach toward graded identity and graded membership in set theory, Fuzzy Sets Syst 161(3):311–346
Jain R (1977) A procedure for multi-aspect decision making using fuzzy sets. Internat J Syst Sci 8:1–7
Lotan T, Koutsopoulos HN (1993) Route choice in the presence of information using concepts from fuzzy control and approximate reasoning, special issue on applications of fuzzy set theory to transport problems. Transp Plan Technol 17(2):113–126
Nagoor Gani A, Mohamed Assarudeen AN (2012) New operation on triangular fuzzy number for solving fuzzy linear programming problem. Appl Math Sci 6(11):525–532
Smith R (2014) Aristotle's logic. In: Zalta (ed) The stanford encyclopedia of philosophy Edward, Springer, Berlin
Zadeh LA (1965) Fuzzy Sets. Inf Control 8:338–353
Zadeh LA (1973) The concept of a linguistic variable and its application to approximate reasoning. Memorandum ERL-M 411, Berkeley
Zadeh LA (1996) Fuzzy logic = computing with words, IEEE Trans Fuzzy Syst 4(2):103–111
Zimmermann HJ (1987) Fuzzy set, decision making and expert system. Kluwer, Boston

QFD Case Histories 5

Abstract
Reading Quality Function Deployment case studies can be useful for understanding how professionals applied QFD and what the specific findings were. In this chapter, we take a look at eight real QFD projects, four in detail and four others briefly. However, imaginary names have been used for the companies involved. The Geotherm case study focuses on a small company that sells geothermal heat pumps, it is a longitudinal study, and in fact QFD was applied on two occasions. Wepartner is a large association of small and artisan companies, that applied QFD to study a territory and its potentialities. Citymove is a public transport company and the aim in that case was to understand who the customer was and how the service offered could be redefined. Elight shows the case of a lighting company which was going to plan the new product line. Paint is a paints producer that deployed QFD to design its production process better. Insure is a telephone and web based insurance company and the case study concerns the organisational planning of the Insure call centre. Mobile 1 is a project our team managed some time ago concerning new product planning for a mobile phone. Stone provides services to marble workers. Our target was to study the needs of registered companies and increase service performance through QFD. Each case has its own peculiarities that will be discussed. The reader needs a certain amount of knowledge about QFD, so I recommend reading Chaps. 2, 3 and 4 beforehand.

5.1 Introduction

Reading Quality Function Deployment case studies can be useful for understanding how professionals applied QFD and what the specific findings were.

In this chapter, we take a look at eight real QFD projects, four in detail and four others briefly. However, imaginary names have been used for the companies involved.

D. Maritan, *Practical Manual of Quality Function Deployment*,
DOI 10.1007/978-3-319-08521-0_5

The Geotherm case study focuses on a small company that sells geothermal heat pumps, it is a longitudinal study, in fact QFD was applied on two occasions.

Wepartner is a large association of small and artisan companies, that applied QFD to study a territory and its potentialities.

Citymove is a public transport company and the aim in that case was to understand who the customer was and how the service offered could be redefined.

Elight shows the case of a lighting company which was going to plan the new product line.

Paint is a paints producer that deployed QFD to design its production process better.

Insure is a telephone and web based insurance company and the case study concerns the organisational planning of the Insure call centre.

Mobile 1 is a project our team managed some time ago, concerning new product planning for a mobile phone.

Stone provides services to marble workers. Our target was to study the needs of registered companies and increase service performance through QFD.

Each case has its own peculiarities that will be discussed.

Table 5.1 illustrates the company characteristics for the detailed projects. As we will see, company dimensions, field of activity and project costs can vary a lot.

The reader needs a certain amount of knowledge about QFD prior to continuing with this chapter, so I recommend reading Chaps. 2, 3 and 4 beforehand.

5.2 Company 1. Geotherm. A Longitudinal Case Study

Heat pumps are one of the fastest growing applications of renewable energy in the world (Maritan 2010). During winter air to water heat pumps and ground-source (geothermal) heat pumps (GSHPs) extract solar energy heat from the outside air (air to water heat pumps) or from the ground (water to water and water to air heat pumps). During summer, these systems expel excess heat from the house to the outside air or to the ground which is like an unlimited battery recharged by Sun. In fact about half of the solar energy that reaches our Planet every day is stored in the ground.

Air to water or air to air heat pumps usually need equipment outside the building, consisting of a compressor, a refrigerant circuit (a liquid that evaporates at very low temperatures), one or two heat exchangers and a fan, that allows for the exchange of heat (energy) with the surrounding air.

Ground source heat pumps extract energy from or expel it back to the ground which has a relatively constant temperature all year round. This is the main difference between GSHPs and Air to water or air to air heat pumps. The performance of these latter types varies a lot during the day and according to the seasons because of the variations in outside air temperature.

Geothermal heat pumps systems can be subdivided into three parts:

Table 5.1 Case study characteristics: market, dimensions, lead time, costs, peculiarities

	Sector	Business	Context	Empl.	Lead time	QFD cost	Phase reached	Largest matrix	Peculiarities
1. Geotherm	Renewable energy	B to C	Strategy	4	3 months	2,000 Euros	(c) HoQ bottleneck	HoQ 11 × 18	Small business
2. Wepartner	Association of small companies	B to B	Services	200	12 months	125,000 Euros	(d) Functions and mechanisms	HoQ 38 × 54	Area marketing, fuzzy QFD
3. Citymove	Transport	B to C	Services	500	5 months	4,000 Euros	(c) HoQ bottleneck	HoQ 61 × 75	Service bus
4. Elight	Lighting producer	B to B	Product line	50	15 months	31,000 Euros	(d) HoQ bottleneck	HoQ 47 × 60	Small company, fuzzy QFD

- the internal distribution system, made up of heating (and eventually cooling) under floor heating (or cooling) system, radiators, consoles, air ducts, and hot water production;
- the refrigerant circuit inside the heat pump, made up of a compressor, copper pipes, filter and expansion valve and two heat exchangers, one on the ground side, the other on the building side;
- the ground loop.

The ground loop is generally consists of polyethylene pipes, buried in the ground either in horizontal trenches (from 1 to 1.5 m deep) or vertical boreholes (from 50 to 250 m deep). A mixture of water and non toxic antifreeze circulates inside the pipes and exchanges heat with the soil. Open loops exchange energy directly with the water from an aquifer: this water is pumped to the surface then exchanges heat with the heat pump through a compact plate heat exchanger. The water then returns to the same or another aquifer, through a second borehole, or a sub-irrigation horizontal line.

Geotherm was founded in the year 2000 after some years of groundwork and planning. The inspiration for the firm came by chance from a short article with the description of heating and cooling systems in the US, that were "plugged into the earth". Today Geotherm is a small technical and commercial company that sells heat pumps and ground source related equipment. It has a group of selected suppliers from Sweden, Switzerland, Italy and the USA.

Geotherm conducted two QFD surveys, one in 2005–2006 and another in 2013–2014. I will identify them as Period 1 and Period 2.

In this case study we can compare how the Geotherm team deployed QFD in two different time periods.

In Period 1 the team was very small: the company owner and his father in law, a technician. Meetings usually took place after dinner in front of a fire.

In Period 2 the QFD group was made up of 4 people, two from marketing and two from the technical office.

5.2.1 Market and Strategy Analysis

By the year 2005, people had become increasingly interested in geothermal heat pump technology, consequently more and more installations had been set up. Geotherm actively spread information about this type of installation, about the high levels of efficiency, reliability and performance and respect for the environment. These were still pioneer years: customers needed to be informed of the existence of geothermal technology. But once the customer was convinced, costs were secondary.

However in Period 2 (2012–2014) the market situation was different. The new building market crisis involved renewable energies too, causing a notable drop for the photovoltaic industry and a similar one in the heating sector. The government introduced unclear laws and a decentralised bureaucratic system which made obtaining permission to drill complicated. In Period 2 customers were confused, a lot of plumbers and firms suggested using heat pump systems, most of them air to water. Therefore customers were looking for lower prices and wrongly though, mostly from internet search, they knew everything about heat pumps. Geotherm had to face new challenges (Table 5.2).

In Period 1 the reason for using Quality Function Deployment in Geotherm was the need to clearly define the strategic priorities and the main performances of both product and services, within a set period characterized by:

- commercial instability, sales network not developed;
- network with partners and other professions (i.e. drillers, excavators, building firms, electricians, refrigerant specialists and plumbers) not developed.

Main targets of Period 1 QFD project were:

- to deploy the three strategic matrices;
- to create Preplan;
- to deploy products and service characteristics House of Quality.

During Period 2, the reason for using Quality Function Deployment in Geotherm was the need to understand the changing customer better: "How should we optimise service characteristics to reduce costs and do a better job by focusing on customer expectations?"

For example, existing customer satisfaction was very high but not used for marketing purposes. In Period 2 the Geotherm sales network was not optimally developed and needed to be improved. Main suppliers were selected and some of them were strong business partners for Geotherm.

As regards the QFD team, in Period 1 two people were involved, the owner of Geotherm and an external expert. In Period 2 four people were involved:

- company owner;
- two people from marketing and sales;
- one person from the technical office.

Table 5.2 Main GSHPs market characteristics for Period 2005–2006 and 2012–2014

Period 1	Period 2
Growing building industry	Building industry crisis
Customers do not know GSHPs	Customers think they know, are confused
Few competitors	A lot of competitors, mostly air-to water
Price is a minor problem	Price is a problem
Few regulations	Several unclear regulations

The company owner, who knew about QFD methodology, trained the others during some initial meetings and guided the group throughout the necessary phases.

Let's start with Analytic Hierarchy Process Matrix 1.

The output of this matrix is the priority of the strategic targets. We can compare results for the two periods.

In Period 1 the targets identified, listed in order of priority were:

- Synergy with strategic suppliers;
- Optimise the choice of making or buying;
- Increase the number of installed units;
- Increase turnover;
- Consolidate a market leader position.

In Period 1, Geotherm had a serious problem: that of securing strategic suppliers.

Two main key components of Geotherm systems, the heat pump and the ground heat exchangers, had to be secured in the supply chain. This was followed by optimizing the "make or buy" choices and increasing the number of units installed. The fifth target has the lowest priority and this means that Geotherm intention was to organise a "network of firms" rather than focus on its position as a leader in the sector.

In Period 2 the targets are very similar, except the search for new market opportunities. They are shown in order of importance:

- Increase turnover;
- Increase the number of installed systems;
- Look for new market opportunities;
- Consolidate the market share;
- Synergy with strategic suppliers;

As you can see the priority of these targets has changed compared to Period 1. In Period 2 the most important are "increase the turnover", (weight 53 %), and the number of systems installed. Synergy with suppliers is the last target. It is interesting the third place "Look for market opportunities": the idea is to develop technologies liked to heat pumps and strengthen the good I.T. DNA make up of Geotherm.

Let's consider core competencies analysis. Here are the lists for Period 1 and Period 2, presented in order of priority (from highest to lowest).

Period 1

- Design and technical know how;
- Flexibility to customer's requests;
- Reliability during installation process;
- Lean organisational structure.

Period 2

- High IT competencies;
- Ability to work in niche markets;
- Broad technical competency;
- We are able to find answers to customer questions;
- Orientation towards reaching good level of customer satisfaction.

Geotherm's core competencies priority (matrix 2 of the framework) changed from Period 1 to Period 2.

From technical know how and flexibility they changed to Information Technology competencies and ability to work in niche markets. Reliability disappeared and technical know how is in third place in the Period 2 priority list: these aspects mean that reliability is probably still important but has become a must-be quality, in order to prevent costs and dissatisfaction in after-sales management. Reliability, in fact, is part and parcel of the company's ability to work in niche markets. Similarly Geotherm know how is still present as main core competency but not in first position. IT competencies are considered decisive in meeting strategic targets also because in Period 1 Geotherm discovered that web marketing was very important.

Finally let's discover what happened to customer priority, shown here in order of priority (from highest to lowest).

Period 1

- Residential end-user;
- Business end-user;
- Public sector;
- Installer/plumber.

Period 2

- Residential end-user;
- Architects/heating engineers;
- Sales representatives;
- Installers/building contractors;
- Business end-user.

In Period 1 two alternatives were open to Geotherm, that of becoming a heating firm within the traditional channels in the sector (special relations with equipment technicians, architects and engineering offices) or concentrating on the end-user. Geotherm felt, and QFD confirmed this, that they had to develop direct contact with end user.

If we compare the customer segmentation of the two periods, we see that domestic end-users still remain the most important. Most of the customers that use Geotherm systems are of this type. Business end-users (offices and industry) moved down the rank to the lowest priority, public sector disappeared, substituted by heating engineers and architects, creating a kind of filter in the relationship between the company and the residential end-user. It is clear that Geotherm wants to increase its sales representative network with domestic end-users the main customers.

5.2.2 Gemba and Preplan

During the Period 1 project, *gemba* interviews were conducted with lead users and few customers. Very little time and hardly any economic resources dedicated to this project did not allow for a detailed analysis. The KJ method was used by only two people, the company owner and a technician, during their free time in a couple of evenings after dinner.

Period 2 *gemba* visits were planned with precision. We concentrated on end-users, followed by architects and engineers and a sales representative. Eight interviewees were involved, we visited the end-users at home and the meeting took place standing in front of their functioning Geotherm geothermal heat pump. Interviewers used a tape recorder.

A sample of voices of the customer/comments from the customers follows.

Interviewee 1: a customer that afterwards became a Geotherm partner. He is an architect. (GSHP horizontal loop)
Place: He was in his office
Interview: Skype conversation

- My inclination towards ecology and energy saving drove me to choose a heating system with geothermal technology.
- Geotherm strong points: they respond very quickly and they have skilled and expert engineers.
- I suggest Geotherm offers a free consulting service giving advice about geothermal systems, solar systems, photovoltaic etc. to end-users.
- Knowledge of the brand in the area is fundamental: I suggest advertising in stores and shopping centres, by mailing, or in newspapers, as websites are just for keen computer users. In my area nearly 70 % of people don't use internet.
- The offer is good, in line with market prices.
- End-users need to be persuaded, they are the ones choosing us, not the installers.
- The installers are often bottlenecks; they don't suggest using geothermal technology.
- There was a breakdown and your after sales-service came out immediately, this kept the customer very satisfied.
- The technical room is nice, very well-engineered.
- The heat pump works well, we noticed good savings.
- The documentation is adequate.
- A payment in instalments system backed by a bank would be a great help in this period of recession.

Interviewee 2: engineer, he worked with our systems but has a competitor's geothermal heat pump (GSHP open loop) in his house.
Place: Technical room
Interview: Direct dialogue

- I found your company on the Internet, then I phoned and I got a very good impression, particularly of you, you seemed to know your stuff more than other people I talked to.
- I chose a different heat pump because it lets me manage everything from a single electronic control unit.
- I am surprised by how much better electricity consumption and the COP (Coefficient of Performance) of your heat pump are, compared to mine.
- I chose a heat pump because I have a well insulated house, and combining it with photovoltaic panels seemed to be a perfect solution.
- No brochures but Internet; a nice website is important with examples, installations you have already done, cost savings, etc.

Interviewee 3: domestic end-user (GSHP horizontal loop)
Place: technical room
Interview: direct dialogue.

- Surfing the web I compared your product with other competitors and I saw that the components you use are considered the best.
- The pictures of compact ground heat exchangers, which no other competitor is offering, was the thing that convinced me.
- One day, when I build my dream house in my dream place I will install the geothermal and compact ground heat exchanger system there too
- A lot of people ask me if it really works, I know that in Northern countries they use this system.
- The 55 % tax deduction was an incentive.
- I had always used a wood-burning stove, both for heating and hot water production. I calculated that I would spend the same, so I decided to change to the geothermal system.
- I found it difficult to convince my plumber to install the system.
- Helpfulness, clear explanations, always willing to give information. You always called me back even just for little things. I have never felt ignored or that you don't care
- I suggest you offer "turnkey systems".

Interviewed 4: domestic end-user (GSHP horizontal loop)
Place: he was at work
Interview: Skype

- I wanted an environment friendly system, so I chose a geothermal heating system.
- I read up about photovoltaic systems, and then I discovered geothermal science. I compared Geotherm with some other companies; your offer was 10 % less.
- Another company had nicer, better leaflets, with fuller information; they also gave me an estimate of electricity consumption over 1 year, you didn't.
- Geotherm strong points are quick response and the competency of the technicians.
- There were some difficulties while I was talking with your partner, but everything was solved when I started talking directly to you.
- The offer was good; it remained the same until the end.
- Your technical service works 1,000 %.
- The technical room is nice; the heat pump is fine, full of electronics.
- The documentation and manual is long enough, if it were any longer, I wouldn't read it.

Using the KJ method (see Chap. 2) the two teams created, respectively, 11 and 19 Demanded Qualities (DQ) in Period 1 and 2.

How did they manage the *questionnaire and the survey*?

In the first case the team sent the questionnaire via fax and email to existing and potential customers and partners. About 50 completed questionnaires were the basis for a raw 5-point Likert assessment for average calculations.

Things were different in Period 2.

The first step taken by the team was to discard the printed option, which would have required a lot of time and high costs: questionnaire design, printing, put it into an envelope, interviewee address selection, sending by post with a pre-paid reply envelope, and then a huge amount of work to transfer the data from paper to a computer spreadsheet. The team was attracted by web survey services, but in this way they also discarded potential customers who do not use internet. A suggestion for the reader is to always consider the two sides of an option, without always being only mad about technologically advanced solutions.

From the various options available the team selected *SurveyMonkey* web service which, similar to others, offers the Plus version at a cost of 25 Euros/month. This version allows you, among other things, to design the questionnaire modules with an intuitive interface, send it to interviewees, collect answers, manage them and create personalised reports.

The questionnaire form can be pasted into a website or a blog, sent by email, sent via social networks. The power of this system is in the mail management: unique web links are automatically created for each email, starting from a mailing list. In this way it is possible to trace single answers from each interviewee. However in this case we preferred filling into be anonymous.

A delicate phase was creating the e-mail address list. As it was important to choose the most useful addresses, SurveyMonkey can manage different folders for a detailed analysis of the data according to how it is collected and should different address lists be used. Addresses can be imported from a .csv file or just copied and pasted into a specific form made for the purpose. Double addresses and invalid addresses are automatically cancelled. Once the address list is ready the sending of e-mails can be planned, with a tailor made text.

We collected all the e-mail addresses from customers over the past few years, so as to have a full list of just potential customers. With Thunderbird tool "Import Export Tolls" we just had to collect all the messages in a folder and export the folder index. In this way we collected 5,700 addresses. 10 min after dispatching the e-mail we already had the first results, ready to be consulted. We closed our survey after 15 days with a feedback of about 13 %. Few answers came from the blog or from social networks, 1–2 %.

All results were entered into an Excel file and the PrePlan was created.

SurveyMonkey offers the possibility of having very detailed reports, divided according to geographical areas or other customised criteria, with very intuitive diagrams which are very useful for presentation of the data later.

It is worth saying a few words about the privacy policy: we should not include people who didn't agree to have their personal data used or passed on to third parties in the address list. SurveyMonkey is certified TRUSTe, they just securely store data and addresses which remain property of the firm creating the survey.

We must also cite the Forms option from Google drive which has its pros and cons:

- pros: simple, free, complete for creating surveys and result analysis (excel sheet);
- cons: very good if incorporated into a website, otherwise it cannot manage the dispatch, it just supplies a link to paste into an e-mail message but this service doesn't manage both e-mail sending and address list like SurveyMonkey does.

If we don't count the time required for the selection of the questions and the learning curve for the interface, we can say that, today, *all the resources needed for a survey like this are a few hours of computer work and the cost of 25 Euros. 500 questionnaires returned back.*

We can now compare the two *Demanded Quality Deployment Charts* and discuss the surveys. Demanded Quality lists are presented as follows in order of priority from Preplan matrix:

Period 1

- Demand for savings;
- Demand for one system for both heating and cooling;
- Demand for optimized space;
- Expectation not to be dependent on fossil fuels;
- To be able to control the system simply;
- To be able to adapt the system to the existing one;
- Feeling relaxed with after sale service;
- Expectation of respect for the environment;
- Comfort in every season;
- "Natural" feeling during cooling;
- Demand for Safety while a heating/cooling system is running.

Period 2

- Respect for the environment;
- Feeling of being in the hands of experts;
- To feel reassured by local testimonials;
- Knowing what I am going to buy;
- No worries after system start up;
- I can call a local company;
- To get an estimate of energy savings;
- Be sure of spending the right amount of money;
- I want to be energy independent;
- Simplicity in system management;
- Request for both cooling and heating;
- To be sure that I bought the best, having evaluated the offers;
- Feeling of state-of-the-art of technology;
- Custom made solutions;
- Need for safety at home;
- No worries during the planning and installation phases;
- I wish to spend as little as possible when I buy the system;
- I wish it to be aesthetically pleased;
- I wish to be different from other homeowners.

Some changes in customer behaviour are evident. From Period 1, where demand was for savings, heating and cooling with the same system and optimised space, customers moved to requests that, in some way give assurance to them and to the next generations. In fact a surprise was the request that the environment must be respected (an assurance for the future) followed by the need of the user not to feel on his own from the planning and installations phases up to after sales, hence the need to have testimonials.

Surprisingly, even during this period of financial crisis, the customer is not interested in spending as little as possible, he is ready to spend the right amount of money.

5.2.3 House of Quality and Bottlenecks

It is interesting to compare the results of the two periods as regards the characteristics of the product/service.

Let's take a look at the two Quality *Characteristic Deployment Charts*, decreasing priority ordered.

Period 1

- Number of hits to internet site;
- Level of personalisation of the installation;
- Selection of suppliers;

- Range of temperature supplied;
- Percentage of financial analyses carried out over total estimates;
- No. of external components of heat pump;
- Courses and specialisation courses (internal) per year;
- dB noise level (decibels) of installation;
- Rate design lead time/installation lead time rate;
- Total space required for installation in "technical room";
- Average time taken for end-user to learn how to use it;
- Clear, simple instructions etc. re the installation;
- Time for first estimate;
- No. of documents that can be downloaded from the web;
- No. of publicity inserts in newspapers and magazines;
- No. of contacts/year with bio architectural associations;
- No. of scientific publications per year;
- No. of Tests before starting installation.

Period 2

- No. of testimonial systems to be visited;
- Energy efficiency of the systems offered;
- No. of partners and sales representatives;
- No. of systems already installed shown on the website;
- Customer reviews;
- No. of social and ecological items of news on the website;
- Personalised system owner's manual;
- Project management procedure shown on the website;
- Safety advantages explained to the user;
- Incentives procedure;
- Website section reserved for partners/designers;
- Max offer lead time.

The two priority trends show a correlation with the market situation.

In Period 1 the building market was growing strongly. A niche market like that of GSHPs was characterised by lower growth.

The first two characteristics concern the *importance of services*: internet and the level of personalization together account for more than 18 % of relative importance. Different performances of the installation fall into the intermediate area and would seem, together with those related to reliability (see for example the position of the item "selection of suppliers"), to be balanced by those relating to services.

Service performances covered more than 70 % of relative importance.

The high level of innovation in Geotherm, which is more apparent than real, in that this technology is already well-established at the global level, could lead one, at first, to think that product performances of the installed appliances and of the heat pumps would be considered critical.

In reality, bottleneck analysis (map of product and service performances based on two variables: weight and difficulty, see Chap. 3) revealed that the technical characteristics of the product were in the part of the map with technical targets considered to be relatively easy to reach. All the bottlenecks were services.

In Period 2 the building market is going through a difficult financial period. Potential customers are confused; they can't seem to rely on anybody.

Characteristics selected by the team are all services.

Four out of five characteristics with higher priority mention building up trust.

Energy efficiency is the fifth element of this group.

The way ahead for Geotherm's next service strategies is clear.

5.2.4 What We Learned

- Longitudinal QFD studies can identify new trends for a company. The Geotherm experience confirms that it is better to plan a QFD analysis every year or at least every 2 years, to anticipate future variations in the demanded qualities priority.
- Surveys conducted using web services significantly lower the costs for this phase. A traditional approach can be used for specific situations.
- Service characteristics in niche markets could be predominant compared to product technical features.
- A QFD project can be developed in case of a limited budget available.

5.3 Company 2. Wepartner. Territorial Marketing with Fuzzy QFD

The Wepartner project takes its name from an association of artisan companies. Wepartner came into being just after Second World War and is made up of about 18,000 associated companies, most of them small companies or sole traders. Wepartner promotes the growth of the member companies, with a counselling role and offering other useful services.

The Wepartner QFD project was designed to increase opportunities for the members offered by the surrounding area (European target no. 2) of about 900 sq km, and including 45 towns and villages with a total of about 200,000 inhabitants. Consolidated small firms, dealing both in traditional and innovative fields, were already operating in this area and craftsmanship constituted the framework of the local economy, not only in the traditional fields such as food, textile, leather garments and wooden products but also in machine design and construction, steel industry, precision appliances, electrical appliances, building and services.

The budget for this project was very high, about 200,000 Euro, and the lead time planned was a year. Initially the Wepartner project was divided in three macro phases:

1. *QFD, Needs and Resources*; during this phase, due to last about 8 months, we created 3 strategical matrices, the gemba analysis to identify the company Needs, the Preplan (fuzzy), the House of Quality matrix (fuzzy), correlation between company needs and territorial resources (characteristics) of the territory and, finally, analysis of the improvement targets and possible bottlenecks while reaching them.
2. *QFD Organisational Analysis*: during this second phase, expected to last 4 months but actually lasted three, our team analysed what the association actually did for its members, its organisational structure (mechanisms) required to perform the different tasks, identifying the suitable ones and the priority order by using fuzzy calculation. Analysis of the mechanism costs was scheduled, but the association decided not to implement it.
3. *Dissemination across the Territory*; the third phase, lasting about 5 months, was directly developed by the association and the external members of QFD were not involved in it. The association activated some start-up services, and carried out a rather good advertising campaign by the member companies. As I will explain, a real link between the QFD results and implementation of phase 3 was missing; there was a lack of feedback, which would have given a result to the analysis.

In addition, phase 2 was only partially executed.

This is the reason why I consider this project as a partial failure, in spite of the huge economic investment.

The QFD team involved a total of eleven people, made up of 4 people from an external engineering company, 2 from Padua University, Italy, and 5 from the Wepartner association. The engineering company advised on the method to follow and implemented the fuzzy calculation software, University carried out a study on the characteristics of the territory, and the Association carried out activities in the field and filled in the matrices, as well as managing the third phase of dissemination.

5.3.1 Fuzzy Strategic Matrices

During a period of about a month, the three strategic matrices (see Chaps. 2 and 4) were created and filled in following several operative meetings with the management.

For the framework of matrix no. 1, 13 strategic targets were identified and later weighted using a fuzzy AHP (priority index with defuzzification method, Chap. 4).

Some targets were:

- Support cooperation between firms;
- Know-how centre for our companies;
- Identify opinion leader customers;
- Maximise the presence of Wepartner in local authorities.

This time, as on several other occasions which followed, we had what we call "*response effect*" (assessment distortion when the number of matrix cells is high). The presence of an external consultant, a specialist and objective guide, proved to be very important as he advised the decisional group about odd situations or contradictions.

The second part of the analysis focused on strategic competency priority concerning their ability to achieve the fixed targets. Nine "Internal competencies" were identified, among them:

- Capillary presence across the territory;
- Potential offer of a broad-spectrum diversified services;
- Being recognised by local authorities;
- Being able to stay in direct contact with companies;
- Financial support for articulated projects.

They were compared to different targets and were later weighted.

We found that "being able to stay in direct contact with companies" was the most important competency for reaching the fixed targets.

Concerning the evaluation priority index of the customer segments (third strategic matrix), the companies were subdivided according to criteria like dimensions (number of employees or turnover), production field or area.

Division according to territorial area was chosen, considering 8 areas identified by following the suggestions of sales people from Wepartner.

These areas were later linked to the internal competencies.

5.3.2 Gemba Analysis and Fuzzy Preplan

Gemba

Gemba analyses were made on a sample of 14 companies, among them small member-companies, non-associate companies and potential start ups. Added to these were 5 interviews with opinion leaders, mainly directors of local bank branches, mayors and former mayors of the towns in the area, and journalists who knew the territory well. Interviews lasted about 1 h each.

The interviews were held in the company offices or in the Wepartner premises. The services the association granted were actually delivered in the Wepartner premises or by the companies. Each company owner-manager spent most of his time in these premises; therefore this was the perfect place for a *gemba* interview (see Chap. 2).

The presence of the association employees was also very useful as it created a very familiar, safe atmosphere.

The 657 "*Voice of the Customer*" sentences, collected in the Voice of the Customer Table (VOCT), were used as a base for implementation of the KJ method.

Using the KJ method, 38 company Requests were defined (Demanded Quality DQ).

The DQs deal with different subjects, from aspects related to the need to conserve the local identity, to themes linked with the company business, concerning product or service innovation

We can make a short list of them as examples:

- feeling rooted in the territory where we work;
- wish to be listened to;
- wish to be part of a group having a strong identity;
- need to be involved in the decision concerning problems of the sector for which we work;
- wish to have more cooperation between the various artisans;
- wish to have fair competition among the firms;
- feeling sure of being paid by customers;
- wish to have a good lifestyle;
- need to make the company bigger;
- need to operate in new markets;
- need to find financing systems easily;
- wish to work adhering to all norms and laws;
- need to find specialised personnel;
- possibility of moving the company to other territories/areas according to business requirements;
- wish to know the technical problems concerning my work.

As we saw in Chap. 2 the Demanded Quality Deployment Chart is the base on which to build a closed answer *questionnaire* to submit to customers or potential customers. In this case the panel included about 400 companies and the two main questions were:

- What order of importance would you give to these sentences (DQ) in order to attract investors' interest in the area?
- How far are these demands satisfied in this territory?

The questionnaire had a *9 point Likert scale*, according to the fuzzy model presented in Chap. 4, and was delivered in paper format, when Wepartner employees made a specific visit to the companies. This investigation process constituted a huge cost for the association, which was not foreseen in the project budget.

After delivery of the questionnaire, the consultants took over the work and entered the data in a data-sheet, followed by the Preplan.

Preplan fuzzy

As seen in Chap. 2, the Preplan aims to determine the customer request priority. This is a relative (percentage) weight that is the combination (product) of degree of importance, target of assessment and sales points; these last were identified by considering the communication strategies of Wepartner.

To summarise, by the following terms, we mean:

- importance = importance of increasing investors' interest in the territory;
- assessment = how much each single Requirement was satisfied considering the area where the company operated;
- sale points = present communication strategies across the territory.

Fuzzy mathematics has been illustrated in Chap. 4. If you refer to fuzzy Preplan calculations, in this case, Quality Plan (QP) was set starting from the assessment of Wepartner, without competitors. If we refer to Sect. 4.4.2, we remember that when available assessment is only for our product/service, Triangular Fuzzy Number TFN QP (a^, b^, c^) elements can be set as follows.

Let TFN (a′, b′, c′) be the average assessment of each Demanded Quality. TFN QP vertices are:

$$b^{\wedge} = c\prime$$
$$a^{\wedge} = a\prime$$
$$c^{\wedge} = c\prime + 1 \quad \text{if} \;\; c\prime \leq 8$$
$$c^{\wedge} = 9 \qquad \text{if} \;\; c\prime > 8$$

In this project a particular graph was introduced, that compared Importance (x-axis) to Assessment data (y-axis), defuzzified with MOM (mean of maxima) method. In this way we were able to study the gaps between level of satisfaction regarding the territory and the importance of the company requests.

Later, sales points were identified and fuzzy priority was calculated with a *fuzzy maximising and minimising set method.*

The results were interesting: assessments were not positive for a lot of DQ. Several were only just sufficient and for a lot of them the company assessments were totally insufficient. We grouped clusters of similar ranking. For example self fulfilment, self autonomy and a job in a pleasant urban environment were needs given a high priority by the interviewees.

We defined the bottleneck DQ as DQ with high importance and negative assessment. This was a particular *variation* on the framework presented in this book. “Relationship” needs were the biggest bottleneck cluster; these included fair competition, not feeling on your own in the territory, transport speed, less bureaucracy and involvement in the political decisions about the local area. On one hand it gave us a picture of the cultural limits of those small companies with little interest in innovation and service technology, while on the other hand it showed their humanistic and positive inclination towards better relationships between people.

Let's go to the *DQ priority*.

We used the algorithm presented in Chap. 4. We developed a spreadsheet program, that allowed us to manage the minimizing and maximizing set for the DQ fuzzy normalised weights and obtain the DQ priority.

Considering the decreasing trend histogram we could analyse the highest score Requests, and we noticed the validation of a macro-trend, which could be defined as the need to create a network, made up of infrastructures ("need to create a good communication network") and relationship between companies operating on the territory ("wish for cooperation between artisans", "wish to interact without difficulties with qualified operators of the same production field", "wish not to feel isolated"). "Need to make the company bigger" had a low score, and this indicated the wish of to overcome the concept of small-medium company and concentrate on innovation, quality and creation of a co-operation system, helping to make the entire group of companies competitive on the market.

5.3.3 Fuzzy House of Quality

At the same time as sending out the questionnaire, requiring 3 months to collect a sufficient amount of answers, the project team concentrated on identification of the measurable characteristics of the territory: our work was being transferred to the "technical" area.

A list of characteristics was drawn up, including demographic, economic and social resources of the territory. It was very demanding work, lasting more than four and a half months, owing to the overlapping of the project team other engagements.

As for the customer requests, these characteristics were also defined using technical sessions with KJ type brainstorming; 54 territory characteristics were identified, divided into seven classes according to their territorial aspect:

- demographic and social indicators (i.e. population density, number of immigrants, etc.);
- services to firms (i.e. exhibitions, consultant service);
- education, research and innovation (i.e. colleges, training institutes);
- production system (i.e. growing rate of the companies, number of artisan companies);
- infrastructure (i.e. road network, railway network, internet availability);
- natural resources (i.e. areas with environmental constraints, historical sites);
- quality of life (i.e. kindergartens, water and air quality, average cost of living).

Characteristics and territory resource measurement was very demanding. We had difficulties in defining parameters characterising the resource and in finding out the relative values.

For each characteristic we prepared a form, divided into different sections:

- Characteristic description;
- Analysis and measurement method;
- Description of characteristic evolution.

Diagrams and concept maps were used. Each form was made up of 1 or 2 A4 size sheets.

At the same time as the territory Resources measurement process, characteristics were written in the House of Quality and we started an analysis of the 2,052 links (38 Requests × 54 Characteristics).

As for other phases of the project, fuzzy calculation was used to find out the preference order of the characteristics.

Members of the Engineering group and University students were *thrilled* with the idea of using calculation software to manage such a huge matrix, where the number of cells was tripled to 6,000 because of the use of triangular fuzzy numbers.

However, soon their enthusiasm was replaced by the difficulties of filling in the matrix, and by the repetitive evaluation activity.

In order to somehow make the work lighter we had to create two groups who could fill in the requests rows simultaneously. Response effect and careless choice of the correlation symbol were always a threat. Boredom, tiredness and demotivation accompanied this phase.

And now, even after a few years, the feeling still remains.

We could work for only 3 or 4 h each day for a couple of weeks and several breaks and stops were needed. None of the Wepartner association wanted to help during these tiresome sessions.

For the calculations, by referring to Chap. 4 model and using defuzzification with maximizing and minimizing set, we were able to identify the priority order of the characteristics.

Let's quickly have a look at the results.

We noticed that the most important resources were placed in the "company services" and "infrastructure"; "cooperative activity" had a clear priority, and was to be read as the need to rapidly implement neighbourhood company networking, one of the strategies for action the project had identified.

In second position was "internet availability". Growing interest in this means of communication was evident and was the most important item from the analysis of infrastructure. The road and railway network was ranked much lower than "internet availability" with about half its weight.

A number of other resources were ranked in an interesting position, all of them giving innovation a central role: the presence of "Training centres", "R&D activity" and "High tech firms". Remaining on the "upper" side of the histogram, we identified a number of resources which help to give the territory a high level quality of life: "Hospitals", "Shopping centres", "Sports and recreational centres", "Kindergartens".

The Territory Resource Target analysis phase took about 3 months of work, not counting the opinion leader interviews we conducted at the beginning.

In order to avoid technical bottlenecks it was important to identify feasible technical targets and values to achieve in the future regarding all the different characteristics.

Each characteristic was ranked (5 point Likert scale) according to:

- estimated technical difficulty for achieving the identified target;
- estimated organisational difficulty for achieving the identified target.

We should bear in mind that a characteristic becomes a bottleneck if:

1. it has a high score of the priority index;
2. the difficulty in reaching the improvement target is very high.

In order to identify the target a territorial benchmarking analysis was carried out between:

- region;
- provinces;
- the area concerned by the project.

Targets were defined:

- evaluating the importance of the resources and services offered by the territory;
- considering the level of the characteristics existing in the benchmarking areas of reference;
- deciding if the target values were realistically reachable;
- keeping in mind that the target has to be reached during a short medium period of time: for Wepartner, a period of 3 years from the end of the analysis was fixed.

What were the substantial differences in the technical and organisational evaluation?

The main difficulty was the presence of both norms and organisational problems making it difficult to reach the target of the project.

Once the targets and difficulties had been identified, we created the usual bottlenecks diagram (see Chap. 3). Without analysing each detail, we found that the "high priority—easy" area was empty, the effort required from the social-economic organisations was very high. What was cooperation network needed?

5.3.4 QFD Organisational Analysis

With reference to the model illustrated in Chap. 3, we can introduce two matrices following the HoQ:

- Request–Function matrix;
- Function–Mechanism matrix.

Let's consider *the Customer Request—Function* matrix.

During a few working sessions the QFD team identified a list of Functions, becoming in this case the various activities Wepartner carries out for its member companies.

Below follows the list of Wepartner functions, in a fuzzy decreasing order of priority:

- create a continuous monitoring system of the degree of satisfaction felt by the artisans about the services given;
- improve and renew the service range offered by Wepartner;
- analyse the training requirements of the associate companies;
- establish a front office to explain the new services offered by Wepartner;
- improve communication between branches and the central site;
- inform member companies about meetings, events and other initiatives;
- create a continuous monitoring system of the services quality;
- involve and stimulate public authorities in favour of the artisan companies;
- inform the artisan companies about services offered by the local branches;
- improve and advertise the relationship between artisan companies and schools;
- organise meetings to explain all the opportunities and services offered by Wepartner to the artisan companies;
- help artisan companies in dealing with foreign customer/suppliers;
- help attend/organise shows and exhibitions;
- prepare category meetings stimulating the presence of the associate companies;
- organise meetings between artisan companies establishing common projects;
- help artisan companies find out about and participate in public tenders;
- advertising services as well as to the already well-known accountancy service;
- keep associates updated about norms-procedures-required formalities;
- organise training;
- start or improve relationship with banks of financial partners;
- establish a system to collect and supply relevant data concerning the sector.

The result was that the satisfaction level of the associate firms had to be improved, together with the services offered.

Function–mechanism matrix assessed the organisational mechanisms of Wepartner in carrying out the action planned for that territory.

Let's see the mechanisms in decreasing priority.

- Department for company activities (credit, training, etc.);
- Office for district authorities;
- Office for social groups (women, young people, seniors, etc.);
- Call centre;
- Office for the development of company business;
- Wepartner services office;

- Environment/safety/quality office;
- Office representing companies in schools, colleges or university;
- Wepartner export department;
- Wepartner IT office;
- Accountancy/salary department;
- Credit limits office.

5.3.5 What We Learned

Wepartner is a very interesting case to analyse, mostly because, even though there was a lot of specific QFD work with a large use of fuzzy mathematics, the investigation was not then followed by an implementation phase across the territory.

Using a metaphor, we could say that we built a beautiful building not allowing people to visit/enter it.

This case confirms the statement "less is better".

Here are some considerations about the case:

- Too much theory, too little practice. A poor relationship between team members, consultants and customers leads to partial results, not from a numerical point of view, but concerning the implementation of concrete action.
- Large calculations lead to small results. Huge matrices, a long time required to carry out activities, response effect, resource wastage, and monotonous work, are all elements that significantly reduced the creativity of working in a group.
- High survey costs must be avoided. Employees time spent on the project must be carefully recorded. Internet should be used as the main tool for the survey.
- It is possible to create a bottleneck diagram, similar to the one we saw in Chap. 4 referring to characteristics, including the customer requests, and using two variables, Importance and Assessment, taken from the QFD questionnaire. Here we can identify DQ bottlenecks, with high average importance and negative average assessment.
- The two key questions of the QFD questionnaire, which we asked during the Customer analysis phase, acquire two different structures according to what we mean as Customer, Importance and Assessment. We can say the same for Request, Characteristics and Mechanisms: these words change their specific meaning according to the context. It is very important for the team to clarify this meaning.
- For services, "functions" correspond to "activities" performed during service delivery and mechanisms reflect the internal "organisation", i.e. who performs the activity.

5.4 Company 3. Citymove. Planning Services for Public Transport

Can QFD be useful when planning a new bus service?

Some years ago I was asked to implement a market analysis project for a public transport company. The company managers were worried about a possible loss of its local monopoly because of free market competition in the near future. New research was needed on who exactly used the bus service so as to understand how the service offered could be redefined.

It was a good opportunity to apply QFD.

At that time Citymove was a mixed company with mainly public assets, about 500 employees, most of them drivers, and about 250 buses. Every day, Citymove carried out between 2,000 and 2,700 runs transporting almost 35 million passengers over a total of nearly 9 million km of routes every year.

The QFD team was made up of the sales department manager, the customer complaints department manager, an external consultant and a student from the University of Padua, Italy. We decided to include other experts from Citymove if necessary (the QFD team is often made up of a constant core and other specialists on call).

The Citymove QFD was developed over a period of about 6 months and focused on analysis of the customer and on the definition of service Characteristics.

5.4.1 Strategic Matrices

Targets, competencies and customer segments were identified during some working group meetings. Customer segmentation was the most interesting phase.

If we take a look at the targets list, we recognize some sentences, indicating the existence of a sub-layer of targets, such as turnover, which all companies aim to increase:

- increase safety;
- increase income;
- increase passenger numbers;
- reduce ticket payment evasion;
- increase competitiveness with private bus service;
- turn clients into loyal customers;
- improve the brand image.

The seven targets were compared pair by pair using the AHP method in order to give them a priority. The list above is in decreasing order of importance.

Seven internal competencies were identified:

- investment capability;
- competency in transport planning;
- number of vehicles;
- quality of management;
- quality of front office service;
- knowledge of end-users;
- institutional relations (community, authorities, associations, ...).

The target-competency matrix is the same as presented in Chap. 2 and in the previous case studies. The most relevant competencies were the financial resources available for investment, technical know-how and the bus fleet, each of them with a priority of about 20 %.

Customer analysis is the area the Citymove managers focused on the most, in order to completely re-define customer segmentation.

The city bus user can choose from many other means of transport: bike, mo-toscooter, car, taxi, and obviously, also to go on foot. A comparison with these other means of transport was rejected as they are not homogeneous elements to compare. It was decided to focus on the "Whats" and on customer expectations rather than on the "Hows". The existing customer segmentation in Citymove was based on seven Characteristics, i.e.:

- personal data (gender, age, job, residence, citizenship);
- physical (disabled or not);
- sociological (revenue, likes, belonging to groups);
- type of use (occasional, season ticket);
- service used (urban, suburban, minibus, ...);
- type of use (work, leisure, study, visit).

This classification was too complicated and unsuitable to be inserted into a QFD matrix, the third type, that could be of use for *gemba* interviews.

A new 5W1H chart (Customer Segment Table, CST) was created, indicating occupations of customers. Eight main categories were identified:

- retired or disabled;
- housewife;
- secondary school student;
- university student;
- employed;
- tourist;
- non-European.

Having created a CST we then added the State Transition Diagram (STD, see Chap. 2), illustrating the behaviour of users who travelled on the Citymove service The STD begins with the need for transport and then passes on to finding a bus route, finding the bus stop, buying a ticket, checking the timetable, waiting, catching the bus and then the return trip. Each statement showed how the customer has different needs that require different responses. The table was very useful during the KJ session in helping us identify the Demanded Qualities.

At this point two matrices were created, one for the link between competencies and customer segments and the other between targets and customer segments. Both calculation results highlighted three main categories:

- university students;
- retired people;
- employed people.

These three categories was included in the *gemba* interviews.

5.4.2 Gemba and Preplan

Which location was to be "the place where the true knowledge was"?

An area in front of the railway station was chosen, at nine o'clock in the morning after the peak rush hour.

Ten interviews were collected, lots of them recorded so as to gather as much raw data as possible. An interview form was used to note down service problems together with suggested solutions, pictures and service features.

Interviews lasted a maximum of 10 min because most of the people had to catch a bus. Sometimes interviewers caught the bus as well in order to continue interviewing or to get more information from the user.

Meeting n.9
Customer meeting Table
ID end user and features: University student, 26, F
Date and place of interview: Railway station
Weather: Variable
Seen Product Characteristics: Bus, bus stop
Interviewer: John Parker

Pointing out problems and opportunities

- A sticker to renew your season ticket is a good idea rather than the electronic pass used in other cities.
- Very few inspectors actually come on to check, I saw just a few.
- Some drivers brake suddenly, this morning I nearly fell, luckily there were no old people standing; they should pay more attention, and be more careful, but often they have to brake because there are children crossing the road.
- I have only ever travelled by bus in the evening twice, I was afraid both on the bus and at the bus stop; some immigrant passengers were laughing at me. Once I

caught the bus alone at 9:30, the other time with a friend of mine at 11:30, both times I was afraid.

- On 1st May buses run everywhere, except to… I came here, no bus running; I had to take a taxi. Luckily I found another person going to the same destination who shared the fare!
- It is nice to listen to music on the bus, but nowadays there is nearly no music, except for some bus routes; why only on them?
- The man who sold me the ticket was very nice.
- I take the bus because I can't walk or bike that far, I have a sore back and I can't sit for long on a bicycle, then I have to go over the flyover. If I can't catch a bus I prefer to walk.

Modification of the above into benefits for the user

- I save time at the desk.
- I do not have to pay constant attention to which direction to take.
- I would feel safer and I would rather travel by bus in the evening too.
- Characteristics of the product/service.
- The student season ticket is well-priced.
- On Sunday there are far too few buses, just one an hour.
- Some extra bus runs are needed, especially in the evening and from 12 to 2 p.m.: sometimes I had to stand right near the door or I missed the bus, and old people find it difficult to travel standing near the door.

Interviewer's observations

- Difficult to stand when the bus is moving.
- Continuously looking around to watch the road and any possible danger.
- The interviewee isn't relaxed, she is always tense.
- She is very careful when stepping off the bus, and careful where she puts her feet.

As well as the ten *gemba* interviews, the team managed to recover data from customer satisfaction research carried out by phone some years before, containing further complaints and suggestions.

The next step was create the *VOCT* (Voice of the Customer Table), containing about 200 raw sentences (raw data) expressed by the customers.

Here are some sentences:

- Some rude drivers stop at red traffic lights but won't open the bus door for you
- I have been told that if I lose my season ticket I can't get a replacement one for free, that way I lose money.
- No complaints about information, except for the day when there was the exhibition when nothing was working.
- That's the way our roads are, they can't perform miracles.

- I could catch a Citynew bus but it is more expensive and I would have to walk anyway.
- The number 18 doesn't run as often as the other buses.
- The service is very good because usually there are a lot of buses except for the 18.
- I take the bus because I always get a headache when I walk along this very polluted street.
- It's too far to walk.
- It's not worth getting a season ticket if I consider Saturdays, Sundays and my days off.
- The bus is a little late but at this time of the day it's understandable.
- When it rains the bus can be 10 min late.
- I have never caught the minibus because I don't know where it stops
- It is good if inspectors come on to check; in… they never do.
- The service is better than in…, the buses are newer and cleaner.
- There are too many stops (there's one every 200–300 m) but we need to meet everyone's needs.
- To improve the service, we should change the traffic layout, and be more restrictive.
- Taking my bike on the bus is too expensive.
- The combined season ticket is well priced.
- It is a little expensive.
- I don't know if there is a direct bus route to the city centre that runs more often.
- I've been waiting here for quarter of an hour.

Below are some comments from the complaints archive.

- Lorries have been using the bus stop for a week now. The area reserved for getting off the bus is now occupied by other vehicles, which makes it very dangerous. Do you think there is something that can be done? Putting up a no parking sign?
- The driver verbally attacked the person because he shouldn't have got on the bus at that bus stop; as we were near the end of the route, he should have caught the bus directly at the terminus.
- The driver did not stop at the bus stop, so a lot of people didn't catch the bus.
- The bus went by earlier because it left the terminus earlier. Because of this one person missed his train.
- The bus cut in front of a cyclist to stop in front of a supermarket; the man on the bicycle had to jump into a flowerbed to save his life.
- The user reports the presence of four inspectors on the bus with just seven passengers.
- People criticize the fact that you have to get on at the front of the bus, while there are signs saying Entrance over the door at the back. This is very misleading, especially for foreigners.

- An empty bus coming from… went by without stopping; the next bus stopped, but it was already full of people.
- The driver talked to a woman the whole way, driving slowly and carelessly. This happens every morning, it's always the same woman but different drivers; this is annoying and dangerous.

Some KJ sessions followed the *gemba* interviews and they allowed our team to make a list of 60 Customer Requests.

The DQs were divided into similar groups and they covered subjects such as the environment, comfort, cheap travel, the ability to inform the customer, relationships between Citymove personnel and the user, safety and speed required for the journey as well as more general DQs. This high number of demanded qualities created some problems in data management.

Let's take a look at a list of them. The reader can see the level of detail a KJ session can lead to.

Respect for the environment and for health

- Wish for less chaotic traffic;
- Awareness of pollution problems;
- Ease in reaching bus stops;
- Ease in getting on/off the bus;
- Ease in standing inside the bus;
- Ease in reaching the ticket machine while the bus is moving.

Comfort

- Feeling comfortable at the bus stop;
- Feeling comfortable inside the bus;
- Feeling comfortable at the help desk;
- No overcrowded buses;
- Need for elbow room inside the bus.

Cheapness

- Wish to save money;
- Wish for a service cost proportional to the distance covered.

Information

- Wish for information about the cheapest ticket or transit pass;
- Wish to be understood while asking for help or information;
- Wish to be understood at the help desk;
- Wish to be understood while asking for information over the phone;
- Wish for clear, visible information at the bus stop;
- Wish for clear information in the bus service offices;

- Wish for simple information on the bus;
- Wish that information at the bus stop is simple to understand;
- Wish that information at the help desk is simple to understand;
- Wish that information inside the bus is simple to understand;
- Wish that the personnel can give good explanations.

Relationship between staff and user

- Wish for drivers to treat everyone the same way;
- Wish for inspectors to treat everyone the same way;
- Wish for operators at the help desk to treat everyone the same way;
- Wish for call centre operators to treat all people in the same way;
- Wish for socialization (catching the bus and going to the centre to meet new people);
- Wish to feel important;
- Wish to be listened;
- Wish for the staff to be discreet (they shouldn't make you look foolish).

Service

- Wish to be independent compared to other means of transport;
- Expectation that all areas of the town get the same service;
- Expectation that on certain bus routes the same service runs both at weekends and on working days;
- Wish for a friendly atmosphere;
- Wish for cleanliness;
- Wish that while waiting at the bus stop you know when the next bus is due to arrive;
- Wish to know, while on the bus, where the next stop is and how long it takes to get there;
- Wish for bus personnel to look tidy.

Safety

- On the bus I want to be protected from any kind of bother (nuisance);
- At the bus stop I want to be protected from any kind of bother;
- I want to be protected from pickpockets;
- I wish to be served immediately at the counter;
- I wish to get to my final destination as quickly as possible;
- I would like to be on time for my appointments;
- I wish to move through the traffic quite fast;
- I wish to find a car park without problems.

In the questionnaire the two key questions were:

1. How important are the following aspects in convincing you using public transport?
2. How far Citymove is able to satisfy these aspects?

Referring to the three identified customer categories, retired people, employed people and university students, the first two were contacted by phone to fix a meeting to fill in the questionnaire with a Citymove employee, while the third group received the questionnaire directly at University.

The collection phase was quite quick, it took about a week, and resulted in about 200 questionnaires being collected, half of them from University students.

The form filling phase was quite complicated for the retired people, who often used to comment on the questions recounting personal experiences and this required a lot of time more than standard. Traditional statistic analysis was applied to the data gathered.

The average values of importance and satisfaction levels were compared, thereby creating a *bi-dimensional diagram "Importance–Assessment"*, similar to the one presented in the Wepartner case study.

The *Preplan* was created bearing in mind the monopoly Citymove has, so the assessment targets followed the rule suggested in Chap. 2 and build a constant increase of +1 for the DQs into the plan.

Sale points were selected by Citymove marketing management.

If we were to comment superficially on the results obtained, we could highlight that the subject of safety was particularly clear: "wish for protection from pick-pocketing on the bus", "wish for safety from car accidents during the journey", "wish for a clean bus" and "wish for safety from the accidents getting on/off" achieved the greatest weights.

The wish for equity and justice also came into this group with the "wish for staff to be able to give explanations" and that "drivers treat everyone the same way", but also, with rather lower values, for the same kind of treatment from the operators at the help desk and the inspectors.

At this point an accurate picture of the customer was made, selecting the most significant segments, defining behaviour during use of service and the expectations were weighted.

It was time to move on to the *Request–Characteristic* matrix, the House of Quality.

5.4.3 House of Quality and Bottlenecks

Two KJ meetings made it possible to identify about 75 characteristics of the Citymove service. They were arranged in a two level structure. See an example of some of them as follows:

Global service

- Punctuality;
- Frequency;
- Passenger overcrowding;
- Integration.

Territory covering

- No. of Bus stop per area;
- No. of Bus per area;

Bus

- Comfort of air conditioned vehicles;
- Cleanliness;
- Noise;
- Reliability;
- Vibration inside the bus.

Travel documents

- Fare;
- Office hours;
- Queues.

Stops

- Bus stops for disabled;
- Distance from the car park;
- Accidents on the bus;
- Pickpockets.

Information

- Timetable;
- Urgent information;
- Rapidity;
- Comprehensibility;
- Press conference.

Driver

- Kindness;
- Careful driving;

- Helpfulness;
- Discernment;
- Proper driving attitude.

Staff at the counter

- Politeness;
- Helpfulness;
- Discernment;
- Understanding;
- Prompt replies.

Safety

- Crashes;
- Incidents inside the bus;
- Theft;
- Harassment.

More than 4,500 links were possible between Requests and Characteristics, using the usual symbols with weights 9, 3, 1 as strong, medium and weak relation, and leaving an empty cell if there was no relation.

The matrix was filled in by identifying the measurement criteria of the Characteristics, the Targets and the difficulty to reach them on a Likert scale.

When we looked at the importance of the Characteristics, we noticed that for Citymove important features were: reducing overcrowded buses, improving regular times, increasing punctuality, increasing the security network, increasing integration with other travel documents (Railways network, etc.).

All these aspects were related to the global service.

But it was also important to reduce the number of instances of harassment on the bus.

Features concerning personnel get less importance (except for the driving ability of the bus drivers), so nearly all of them are positioned in the lower middle part of the diagram.

Once the basic Characteristics were identified, the moment came to find out what difficulties there were in achieving them.

For this purpose we created a Priority–Difficulty graph.

This allowed us to highlight both the priorities for the quality Characteristics based on client needs, and to identify the bottlenecks. These latter are features with a relatively high importance (therefore necessary to satisfy the client), but also with a high level of difficulty of being achieved.

The reader will remember, for example, that features with relatively low importance and a high difficulty are the obvious ones to avoid. On the other hand, features with a high importance and low difficulty need to be given greater consideration, because by implementing them, customer requirements can be satisfied easily.

The most delicate part of the Importance–Difficulty graph were certainly the bottlenecks:

- "Overcrowded buses" firstly, and just one grade lower in importance comes a group of three features;
- "Keeping to a regular timetable";
- "Punctuality";
- "Harassment".

Less important and less difficult but still in a critical position was "reliability of the driver", the Citymove employee with the most direct contact with the Customer. In this particular case it was not possible to say if this characteristic had to be improved or not, as the company Citymove decided to keep the score for this Characteristic confidential.

In the diagram area, where important—easy to implement Characteristics are positioned, two features were placed:

- "A correct attitude to driving, particularly regarding the use of mobile phones while driving";
- "Route information" at the bus stop.

In other words, with a small investment of resources, the company could achieve high customer satisfaction. What was said previously about staff Characteristics is also valid for the first Characteristic. For the second one, the target and the current level are the same therefore no changes are needed.

The central area of the matrix contains "indifferent" Characteristics which are neither essential for the customer nor difficult to realize. They could be taken into consideration if, for example, they turn out to be very positively connected (information can be taken from the HOQ roof) to features with higher priority. In that case a possible improvement would make sense.

5.4.4 What We Learned

- QFD applied to a service company working in a monopoly position; competitors are other means of transport and cannot be involved in the Preplan for benchmarking, otherwise each Demanded Quality in Quality Plan section would require the maximum assessment.
- Diagrams speak a universal language; diagrams like 5W1H are very useful. State Transition Diagram is a powerful tool to describe the state a Customer goes through while using a service.
- Pay attention to methodology strictness. A QFD specialist should lead the team in order for the different phases to be understood and applied accurately.

- QFD is not a Customer Satisfaction analysis carried out by the after-sales department. QFD actively drives Customer Satisfaction, upstream from the Customer, not or not only downstream.
- Overly large matrices cause unnecessary tiredness and the loss of focus on real problems.

5.5 Company 4. Elight. Product Line Planning

Elight is a small Italian company that produces both classic and modern lighting systems, mostly focused on the domestic sector rather than offices. Elight's main markets were Italy, Spain, Greece and the United Kingdom.

Some years ago, at the time of the project, Italy, closely followed by Germany, was one of the biggest producer of lights and lamps (wall, ceiling, desk and floor) and the leading exporter state with a sales turnover of 2,500 million Euros.

This market is divided into two segments:

- products for exterior lighting (streets, gardens and urban areas, sports fields);
- products for interior lighting (modern style, classic, industrial, commercial, modular).

Most of the production was designed for the industrial and commercial sectors, followed by modern style lighting featuring close attention to detail in design.

Low cost production and the expansion of retail chains, like the low cost Swedish furniture stores, were creating turbulence in this market for middle class customers, causing an extreme differentiation in purchasing behaviour, high price and designer label or low price with some styling in the product lines.

We were contacted to support the creation of the new line of products and to help our client gain a better understanding of the customer.

The QFD project was deployed in four phases:

1. Strategy;
2. *Gemba* and questionnaire survey;
3. Preplan;
4. House of Quality and Bottleneck analysis.

The external consultancy budget was fixed at about 26,000 Euros, to which internal resource costs, estimated to be about 5,000 Euros, were added.

Table 5.3 shows the planned Gantt diagram. Lead time for the project was fixed at 15 months. We planned about 50 days of man hours: note that phases three and four overlap as do phases one and two, partially. In addition to these 400 h of consultancy work, we counted about 500 h of Elight employees' work (marketing and sales). We will see later where these activities were concentrated.

Table 5.3 Elight QFD project *planned* Gantt diagram (see Sect. 5.5.5)

Phase	Days	Work hours	Costs	Gantt (months)															
			Euro	1	2	3	4	5	6	7	8	9	10	11	12	13	14	15	16
1	8	60	4,300	x	x	x	x												
2	13	100	6,900			x	x	x	x	x									
3	8	60	4,500							x	x	x	x						
4	19	150	10,00							x	x	x	x	x	x	x	x	x	
Tot.	48	370	25,700																

Fuzzy mathematics, as seen in Chap. 4, was used in the questionnaire design, Preplan and House of Quality matrices.

5.5.1 Phase 1. Strategy

Some preliminary meetings helped top management of Elight to learn the basics of the QFD methodology and to start the Strategy phase, with the usual aim of identifying targets and to prioritize them, of selecting core competencies and customer segments, and thereby deploying the three matrices, that we know, are the start of the framework presented in this book.

15 strategic targets were identified, some of them listed below:

- Purchase costs reduction;
- Creating a modern identity;
- Consolidation of the Elight identity as seen by the customers;
- Maintaining and consolidating the customer perceived quality level;
- Production process optimisation;
- Identification of human resources in Elight strategy;
- Gain/maintain customer loyalty;
- Creating a solid sales network in Europe;
- Gain new market share;
- Periodical product innovation.

Using AHP methodology, a matrix was created, which compared targets in pairs for a total of 225 analysed links.

From analysing the results, it was clear that the focus was towards company sales development, with the main aim of creating a European sales network and of keeping existing clients as loyal customers. Next came product innovation. The lower part of the diagram, concerning the Elight strategic targets priority, was to be read with a "conservative" approach: items such as maintaining the present special order turnover, customer quality perceived level or reduction of purchasing cost had to be frozen as they were at the time.

The second part of the analysis consisted in identifying and calculating the priority for Elight strategic competencies concerning their ability to reach the fixed targets.

11 "Internal competencies" were identified and linked with targets from the first matrix; some of them are:

- Loyal salesmen;
- Being flexible to customer requests;
- Being open to innovation and to the ideas of the young generation;
- Being able to produce in-house;
- Financial solidity;
- Good quality-product ratio;
- Made in Italy product.

It was very important to involve the company salesmen in order to reach the fixed targets.

Being open to innovation followed the sales aspect, as in the AHP matrix. Intrinsic quality of the product as indicated by good price/quality ratio or made in Italy were of secondary importance. This was a clear signal: strategic customer and service oriented competencies were to be reinforced; made in Italy lost its distinctive and attractive appeal.

9 main segments of business Customer were identified for Elight.

Attention to business to business marketing was explained by the sales network type, i.e. retailer shops. Advertising aimed at the end-user was still open to discussion, as Elight wanted to increase brand identity particularly with end-users.

Let's list some:

- Lighting shops;
- Franchising shops or purchasing groups;
- Furniture shops;
- European distributors;
- Novelty shops;
- Home appliance stores;
- Electrical hardware stores or wholesalers.

Analysing the matrix results we noticed that lighting shops constituted the prevailing segment, followed by franchising shops and purchasing groups. European distributors maintained high importance.

5.5.2 Phase 2. Gemba and Questionnaire Survey

Over a period of 5 months we designed and implemented:

- the qualitative *gemba* analysis of the business customer;
- the customer requests deployment to prepare the questionnaire delivery to a panel of shops and companies dealing with lighting in Europe.

The 15 customer companies involved in the *gemba* were based in:

- Italy (five);
- United Kingdom (five);
- Spain (three);
- Greece (two).

The collected data were entered in the *Voice of the Customer Table* (VOCT). 425 sentences were recorded generally by means of a recorder.

We can now look at some raw data, extracted from VOCT:

- Quality/price ratio is important. Elight is doing well because it started producing good working items, nothing particularly special, but at a good price for the end user. Now its mistake is to look for self gratification without taking into consideration the market from a pricing point of view. You can't fool anyone in this sector and there is no territory to conquer.
- It is very easy to copy the style of another lamp, play with it and redesign it more or less similar to the original. Elight can do this very well, although not in a very original way.
- The typical customer for classic lighting products is, on average, about 40 upwards, and looking for products normally for an existing house or one undergoing renovation.
- Lucan has had a very standard ceiling light for the last 20 years, with a steel plate/shade that is a nuisance, as it masks some of the light.
- Those green and orange ceiling lights are unsightly.
- Yesterday a lady told me that she only used low consumption bulbs in her house but she kept talking about halogen lights.
- Hammering advertisement makes a company grow from rags to riches.
- Nowadays having 2.70 m high ceilings is different from the past when they were 3 m high, the suspended lamp doesn't work as an idea anymore.
- Most of our customers own their houses, have a job and live in this area. They are rather wealthy as this area is quite expensive. 75 % of our customers work in London.
- Glass is the essential component of a lamp. In order to evaluate lamp quality we consider glass for 60 % and metal parts for 40 %.
- Young customers buy modern products, but not expensive ones, they look for average prices like Ikea's; small modern plastic lamps lasting for about a year or two that can later be thrown away and replaced with new ones. There are a huge number of customers buying like this.

Using the KJ method, which you will remember consists of brainstorming sessions with the use of sticky notes, 47 customer needs/demands were found, and afterwards these were inserted into the Italian, English and Spanish questionnaires.

Let's see, as an example, some of them:

- Simple, clear design;
- Expectation of satisfying customer requests concerning design;
- Wish to concentrate on few suppliers, able to "sustain" my business with services and products;
- Feeling self confident during sale and after sale;
- Wish to fulfil customer requirements about safety;
- Being able to offer technical innovation to the Customer;
- Wish to have something simple and easy to install, both in the showroom and in the customer's house;
- Wish to optimize the selling time;

- Wish to limit investments;
- Wish to meet requirements about easy to clean products.

It is interesting to see *what happened with the survey.*

Italy. During the summer the questionnaire was delivered by post to 300 destinations some Elight business customers, others not, who had previously been contacted by phone and were afterwards asked to send back the completed questionnaire in at least a week. We collected 60 questionnaires.

Spain. Spanish market, with a very lively lighting market, was carefully analysed: a Spanish translation of the questionnaire was sent to 250 Elight Spanish customers. We gave them plenty of time to send back the questionnaire (deadline 4 months); we received back just 10 % of the questionnaires, a total of 20 forms, even after a great deal of work was done sending reminders, making telephone calls and sending faxes.

5.5.3 Phase 3. Preplan

A Preplan was developed using a fuzzy approach to manage Likert assessments in the questionnaire and a method of maximising/minimising sets was used to find the overall priority of customer requests.

The main result arising from Preplan was that quality, as seen by the end user, is one of the main issues for Elight.

A feeling of self confidence during sales, after sales service, purchasing and delivery phases were among the demanded qualities considered most important by the Spanish market, followed by the Italian market and in the end by the rest of the world. Maintenance and easy to handle repairs for the end user were very important only in Italy.

The need to optimize the selling time was an important issue, as was being helped by easy to explain technical features of the product. Foreign customers were very interested in quickness and efficiency while Italian ones were more interested in passively displaying the product in their shops, instead of having to hard sell it to the potential client.

5.5.4 Phase 4. House of Quality and Bottleneck Analysis

The inter-functional QFD team, starting from customer requests, had to identify the product/service characteristics able to respond to customer expectations.

During some of the meetings that followed our technical consultancy team, made up of 2 engineers and 2 marketing specialists, together with the Elight sales managers, identified 60 Elight service/product range characteristics.

The characteristics ranged from service performance (sales and after sales), to product range features and more specific technical characteristics. Some of them are:

- Warranty;
- Number of products in the range;
- Colours available;
- Number of advertisements;
- Assembly and fitting instructions;
- Average time to answer specific requests from customers;
- Time to change lights/bulbs;
- Catalogue;
- Number of designer label products;
- Number of low energy products;
- Time for delivery.

They were entered in the *Characteristic–Request* matrix (the so called House of Quality) activating the analysis of 2,700 links. The fuzzy weights allowed for the definition of a priority index of the Characteristics. Measurements of characteristics concerning Elight and some competitors were collected; technical improvement targets were identified; for each performance a value scale from 1 to 9 (very easy to very difficult) was established, in connection with:

- Estimated technical difficulty to achieve the final target;
- Estimated organizational difficulty to achieve the defined target.

This point leads us to the second part of the analysis, the bottleneck analysis: a characteristic is a bottleneck when:

- it has a high index of priority;
- reaching the improvement target is very difficult.

Technical benchmarking with two competitor companies was possible thanks to interviews with the sales force and from the analysis of brochures and technical documentation.

5.5.5 Project Lead Time and Relationship Effects

It is interesting to note the actual lead time, in comparison to that planned, which was presented in the introduction:

- Phase 1: 70 h;
- Phase 2: 100 h;
- Phase 3: 75 h;
- Phase 4: 105 h.

Phase 4 lasted a shorter time than planned; we speeded up the work on characteristics, by using pre-programmed spreadsheets and a very well trained team for the QFD KJ sessions.

Elight asked our consultancy to write a report about design and styling issues for that year and the next.

Then we added about another 65 h of work to carry out research into styling in areas like the automotive, furniture and jewellery sectors. Four opinion leaders were interviewed.

How was the employees' work subdivided?

About 40 working hours were dedicated to *gemba* visits, 400 h to questionnaire management and about 50 to Preplan and HoQ each.

If project lead time received positive confirmation and can be considered a point of success for this QFD project, I cannot say the same for the management aspects or the relationships between us and our customer.

If we want to be coherent with QFD philosophy we must study errors.

Our consultancy on QFD is rich with errors made and yet, at the same time, growth coming from those errors. We still continue to make errors.

The main error we, or rather I, made, in this case was to under estimate the effects of relationships.

An Elight internship student was involved following this project.

This person spent every day working in Elight, developing his university thesis and learning about the internal processes of this company.

Top management involved him in several activities, offering an overall view of their way of managing the business. He came to be esteemed by the company owners for his self-denial.

Our team was not present in Elight every day, as the company is physically quite far away from our offices.

Inch by inch, the vision of our team about management, activities deployment and choices in the project seemed not to be completely accepted by this person, who started to take a stand against our proposals. It began with some minor episodes, for example during formal meetings with our customer. He seemed to me to be very competitive.

Relationships, day by day, between me and him became more and more suspicious up to the point when I asked for a meeting with top management. Their behaviour was not clear to us.

We couldn't understand what was happening.

I still remember that on one occasion, one of the company owners talked with me about this problem and said: "It seems that there are two roosters in this madhouse".

Then I understood. Better, I thought to have understood.

I decided to ask that this person not be involved in the successive steps of the project. This was a second error, following the first, that was not to immediately understand that there were some problems in the relationship between me and him.

It would have been better to continue to include him more and more. Involving people is risky but their sense of responsibility grows.

This student was employed by Elight after completing his thesis.
Our customer trusted him and we should have been aware of this earlier.
Customers are sometimes so mysterious.

5.5.6 What We Learned

- Importance of relationships. You could be a guru on QFD methodology but team building, psychology and empathy for all those involved in a project are essential.
- Fuzzy QFD confirmed to be a good alternative to crisp logic. We found, as in every fuzzy project we worked on that there was some "flattening" of the priorities trend, mostly because of the stretching of the TFNs support matrix after matrix.
- In phase 4, Request–Characteristic bottlenecks graph can be divided in two, considering organizational and technical difficulty.
- The questionnaire phase was long, exhausting and expensive. My suggestion is, if possible, to abandon the traditional approach and to go straight to web-based applications. The traditional approach is still to be preferred when direct contact is important for encouraging questionnaire returns.

5.6 Review of Four Other QFD Projects

We will now take a brief look at these four other cases that showed errors, peculiarities and some differences from the framework presented in Chaps. 2 and 3.

5.6.1 Case 5. Paint

Case 5 regards a paint producing company which deployed QFD in order to plan its "process" better.

60 % of its turnover is made up of orders from painters and decorators and it was precisely from this customer segment that QFD deployed the technical and process part of the project.

250 sentences, gleaned from 18 *gemba* interviews, were investigated in KJ sessions.

Due to the scepticism, shown during KJ brainstorming, and the difficulties encountered in abstraction, the employee members of the team found the sentences somewhat trivial.

35 DQs were used in creating the questionnaire.

The survey lead time was quite long (about 4 months).

A new step, as compared to our framework, was that the Preplan matrix was split into three matrices, one for each main customer segment (data from questionnaires was divided into the three segments too): steel painters, wood painters, building painters.

3 HoQ with 44 product characteristics were created.

The problem here was to calculate an overall priority that would summarise the three values calculated. We invented a new matrix, where *we ranked the impact of each characteristic on each specific segment of activity*.

Total priority was the weighted sum of the priorities with the weight calculated using the independent scoring method.

The company was unable to measure the characteristics well so bottleneck analysis became an evaluation of the difficulty "to measure" instead of "to reach the target".

It was too difficult to continue with three parallel studies, so the team decided (a little late) to concentrate on the building painters segment.

10 product functions, like "protect", "resist", "cover", were selected and the two matrices Characteristic–Function and Function–Mechanism were deployed.

Parts and Costs analysis was conducted in the same way.

23 Process phases were correlated to parts.

Let's talk about the critical stages.

It would have been better to abandon the in-depth analysis, split into three customer segments. My suggestion is to select the average or concentrate on the single segment that interests you; the team's target is to move quite quickly, using a clever approach. Product characteristic measurements can sometimes be a problem, just as they were in this case.

Do not worry; go ahead; you will have time the following year to concentrate on characteristic measurements.

Sometimes the parts deployment chart can be problematic. One solution can be to identify a standard product, or another is not to go into too much detail. Remember, your target is the planning not the detailed design, *you are going through the Fuzzy Front End period not the design stage.*

Bear in mind time scheduling too.

Cost analysis was the element of this project that was particularly appreciated by the top management. Of course.

I imagine that in the future *Cost deployment will be more and more interesting and a lot of work still has to be done on QFD Cost deployment.*

5.6.2 Case 6. Insure

Case 6 concerned the organisational planning of the call centre for Insure, a telephone insurance company.

It is a typical application of QFD for service design.

In this project we focused on Function and Mechanism deployment.

QFD strategic matrices identified 8 targets (for example, as for other projects, sales increase, turnover increase, customer satisfaction) and 9 core competencies and 4 client segments (job, age, territorial area, level of education). The team conducted 27 *gemba* interviews from a group of 300 potential interviewees. KJ sessions led to 22 demanded qualities, starting from about 120 raw data.

The survey was conducted traditionally, 5,000 questionnaires were sent out and about 150 of them were returned.

Lead time for the survey was 2 months.

35 characteristics of the service were created from the technical point of view, for example "problem solving time", "average No. of complaints about personnel courtesy", "time kept on hold during the phone call".

No bottleneck analysis was conducted.

The team concentrated on call centre functions (20, among them answering complaints, active consultancy, data insertion, reading customer's mails, listening to the customer, giving information about procedures to follow).

Finally four sections of the office became the mechanisms ("Offers", "Renewal", "After sales service", "Business").

To summarise the difficulties we encountered during the process, all that needs to be said is that the Function–Mechanism matrix was unbalanced on the function side. We ran the risk of not being able to differentiate between the mechanisms.

We tried, unsuccessfully, to investigate Cost deployment, the customer did not trust in our discretion and preferred not to develop this phase but keep it confidential.

5.6.3 Case 7. Mobile 1

This case study is about a project our team managed some time ago, and concerns new product planning for a mobile phone. The team tried to develop strategic targets, core competencies and customer segmentation.

Unfortunately the targets were unfocused because the firm was going through a period of transition and reorganisation.

Core competencies were not identified; we got poor support from managers and employees. The company did a detailed study of customer segments and developed a 5W1H chart for the following segments "students aged 14–21", "students aged 22–33", "employed people aged 20–33", "employed people aged 34–60". *Gemba* visits were made to specific focus groups and student interviews conducted in three towns.

Some other material was added to that acquired from the *gemba* visits, such as internet forums and newspapers. Five KJ sessions allowed for the creation of about 40 Demanded Qualities.

Some of them were not completely abstracted from the product characteristics (for example: "intuitive menu", "easy to read display", "easy to carry", "long lasting battery", "protection from electromagnetic field", "it must be reliable").

As you know this is an error that can create confusion. See Chap. 2 and the KJ section about this.

The *Questionnaire* distribution phase was a little delicate; 250 end users were involved and the questionnaire was distributed in a printed paper version.

240 man hours were needed for this phase.

Note the difference between this case and case 1, Geotherm in Period 2, where a web survey was used.

However, during these 240 h the team was able to reach people, they could not have reached had a web survey been used instead. Benchmarking analysis was conducted on two levels:

- lead users compared product specifications;
- end users gave opinions about the company brands.

The benchmarking rank was formed by the addition of the two assessments (lead and end user) with weights fixed by the company being respectively 0.3 and 0.7.

A specific database was programmed to study and sort the answers into segments

Some limitations in the development of HoQ must be highlighted: technical benchmarking was incomplete because of time limits and the company was more concentrated on results from the Preplan matrix. In any case about 50 characteristics were created, considering that the product was aimed at young people.

From studying this case we must remember that it is important *not to confuse customer requests (DQ) with product characteristics* and that *lead time limits* must be considered carefully in order to finish the matrices the team wants to implement in time.

5.6.4 Case 8. Stone

Over a period of 6 months our team, with the help of a brilliant university student, deployed 50 customer requests, programmed 8,500 Triangular Fuzzy Numbers (TFNs) in spreadsheets, identified 80 service characteristics, 90 service functions and more than 11,000 matrix relationships, one after the other.

We had to upgrade the computers used and resort to using a plotter for the print-outs (Candido 2002).

This was all carried out for the Stone association QFD project. Stone is a department of Wepartner association (see case 2) that provides services to marble workers. Our target was to study the needs of registered companies and increase service performance in the Stone department.

The QFD framework adopted was similar to that presented in this book, ending at matrix 6, with characteristics as rows instead of customer requests.

Therefore using the three strategic matrices, a survey and Preplan were implemented and then HoQ, in order to reach Function deployment. Fuzzy mathematics was used for the first time, so we had to study a lot and program calculations to compare fuzzy priorities.

It is interesting to dwell a little on *the survey and other issues encountered* during the project.

Gemba didn't take very long (8 interviews) but the information gathered was enough to write 160 Voices of the Customer.

From these sentences the QFD team identified the 50 demanded qualities and a questionnaire was created. The only possible approach that could be used for the survey was the traditional one and, in addition, face to face with the interviewee. 130 companies were involved in this phase, half of them not registered with Stone.

There was very little collaboration from them, low interest, and difficulty in getting dates for appointments which all caused a long costly process that lasted for more than 2 months. The reader can imagine what the second problem we faced was. The clue lies in the first lines of this case history: the complexity of calculations and the dimension of matrices.

A total of 24 h were needed for KJ sessions and 22 h to fill the cells in the matrices.

It has to be said that the more detailed the deployment charts are, the easier and faster it is to fill each cell of the matrix because relationships are clearer. Instead, having more columns to prioritise generally means a reduction in the gap between each priority.

And this is not a good thing for a decision maker.

Also in this case, while working on HoQ the team got into a little trouble with technical benchmarking and setting technical targets. We had to invent a new index, instead of "difficulty to reach the target" we assessed "difficulty in measuring the characteristic".

In addition, we encountered hostility from managers several times regarding quality of service measurement most likely as they were probably worried to show any weakness in their area.

Sometimes, wrongly, adults do not consider transparency as a strong point and forget that hiding something is just like an invitation for somebody to search and discover it.

The last issue we encountered in this project was the question: "how do we calculate priority from fuzzy numbers?" We adopted the method proposed in Chap. 4.

To conclude this case and this chapter, I must remind you that *team building* is the first objective.

A harmonious, motivated, skilled team, that sometimes has fun or wants to turn rules upside down, like we had in Stone, is the most powerful tool to deal with easy and difficult tasks.

You would think that it goes without saying but just take a look at what's going on in your and/or friends' workplace and you will see that it doesn't.

References

Candido (2002) Teoria ed applicazioni della metodologia QFD nei servizi. Il caso Unione Provinciale Artigiani di Padova (Theory and applications of QFD method in services. The case of UPA Padua). Dissertation, University of Padua

Maritan D (2010) Geothermal heat pump systems for multi apartment buildings: some case studies in Italy and project management implications. Proceedings world geothermal congress 2010, 25–29 April 2010, Bali, Indonesia

Printed by Printforce, the Netherlands